TRAINING SECRETS OF THE WORLD'S GREATEST FOOTBALLERS

BLOOMSBURY SPORT
Bloomsbury Publishing Plc
50 Bedford Square, London, WC1B 3DP, UK
BLOOMSBURY, BLOOMSBURY SPORT and the Diana logo
are trademarks of Bloomsbury Publishing Plc

First published in Great Britain 2019

For picture credits, see page 240

A catalogue record for this book is available from the British Library

Library of Congress Cataloguing-in-Publication data has been applied for

ISBN: PB: 978-1-4729-4845-8; eBook: 978-1-4729-4846-5

10 9 8 7 6 5 4 3 2 1

Typeset in PMN Caecilia and designed by Austin Taylor
Printed and bound in China by Toppan Leefung Printing

Bloomsbury Publishing Plc makes every effort to ensure that the papers used in the manufacture of our books are natural, recyclable products made from wood grown in well-managed forests. Our manufacturing processes conform to the environmental regulations of the country of origin

To find out more about our authors and books visit www.bloomsbury.com and sign up for our newsletters

Picture below: Tottenham's Hotspur Way training facility comprises 15 pitches, of which four are the preserve of the first team

Picture overleaf: A state-of-the-art Airdome, as used by Southampton (pictured), plus the likes of Chelsea and Ajax

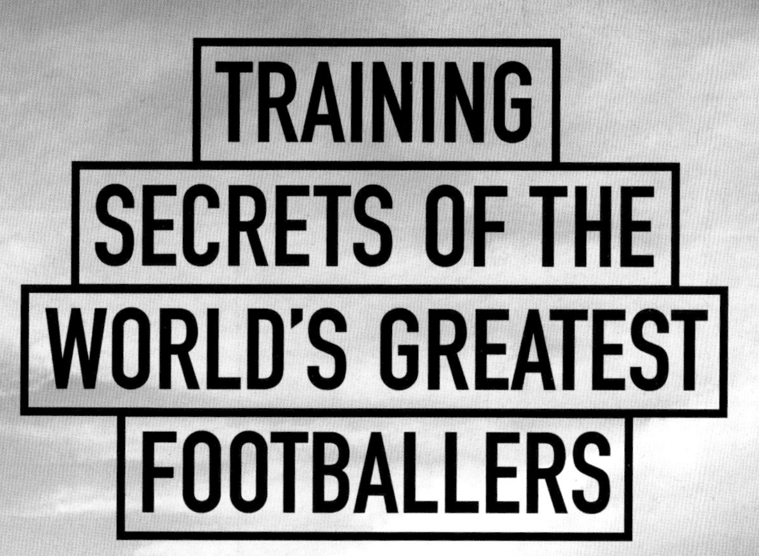

TRAINING SECRETS OF THE WORLD'S GREATEST FOOTBALLERS

HOW SCIENCE IS TRANSFORMING THE MODERN GAME

JAMES WITTS

B L O O M S B U R Y   S P O R T

LONDON • OXFORD • NEW YORK • NEW DELHI • SYDNEY

To Mia & Harry

CONTENTS

FIFA WORLD CUP
RUSSIA 2018
FIFA
VAR ROOM
PORTUGAL
OUTPUT 3

INTRODUCTION

One of the defining features of the 2018 World Cup in Russia happened off the pitch, for this was the tournament of VAR, or the Video Assistant Referee. An evolutionary step from goal-line technology at Brazil 2014, the technical team in Moscow would review incidents from dozens of camera angles and determine whether the referee should reconsider his on-field decision.

'I would say to the fans, players and coaches that it will have an impact, a positive impact,' said FIFA president, Gianni Infantino, before the tournament. 'That is what the results of our studies show. From almost 1,000 live matches that were part of the experiment, the level of the accuracy increased from 93 per cent to 99 per cent. It's almost perfect.'

Croatia fans might argue otherwise. Ten minutes after Ivan Perišić had scored a beautifully executed left-footed equaliser in the final against France, he was judged to have handled the ball from Antoine Griezmann's corner. Initially, the referee, Néstor Pitana, gave a goal kick, only for VAR to alert him that the decision needed to be reviewed on the pitch-side monitor.

Pitana pondered, this former actor still a master of dramatic tension, before signalling with a flourish that he'd changed his mind and pointed to the spot. Griezmann stepped up and slotted the ball into the bottom left corner, scoring the first-ever penalty to be awarded thanks to VAR in a World Cup Final. History had been made ... and the rest is history.

◀ Technology's increasing impact on football was felt most explicitly at the 2018 World Cup with VAR

VAR attracted its fair share of critics; it also overturned incorrect decisions for a 'fairer' World Cup. What's clear is that football, known for its resistance to technical evolution and perpetually reflecting on a rose-tinted past, is now wising up to the benefits of innovation and not just when it comes to refereeing.

Where once a team's support staff comprised the manager, his assistant, a physio and a kit man, now nearly every club features a sports-science department with a far-reaching remit, from developing a greater understanding of how the human body adapts to specific types of training to catering for the specific biomechanical demands of every position. And when I say 'nearly every club', I mean the full spectrum. Here's a snapshot of staff lists from England's League Two: Marcus Flitcroft, match analyst, Exeter City; John Lucas, head of fitness and conditioning, Bury FC; Luke Jelly, head of sports science, Lincoln City.

Then, at the other end of the fiscal rainbow, you have a mega-club like Manchester United, who reportedly employ 300 staff at their Carrington training facility, which in itself is a far cry from the Cliff, the sports ground in Salford that they inhabited from 1938 to 2000. While the Cliff slowly developed over the years, it was still little more than a few training pitches – of variable quality – a gymnasium and a changing room.

Now, United players hone their skills on all-year-round-perfect playing surfaces – they can choose from 12 grass and two synthetic numbers – before ensuring their expensive limbs remain in equally perfect condition thanks to high-tech screening equipment normally reserved for hospitals. We're talking MRI, CT and ultrasound scanners. (Facilities like these don't come cheap, of course – they cost a reported

▼ Néstor Pitana consults with the VAR team in Moscow before awarding France a penalty at the World Cup Final

▶ Even warm-ups have ultimately derived from the work of sports scientists

£60 million – but United soon recouped the lot and more after a £180 million sponsorship deal in 2013 that changed Carrington's name to the AON Training Complex.)

Today's clubs have experts to help players recover faster, run faster and heal faster, with many qualified up to PhD level. Players are trained for their specific positions within a team based on analysis by sports scientists, who tell the coach how many sprints that role demands, over what distance and at what rate of acceleration. This data-led specificity is symptomatic of the age and is light years away from times gone by.

Take this from Dr Neil Phillips, former medical advisor to England's Football Association and the England team during the 1960s and 1970s. 'In the 1950s, lapping around a running track was the sum total of most top players' fitness sessions and many of them never saw a ball from one Saturday to the next, the theory being that it made them hungrier to play in a game when you eventually got to kick a ball. Only in 1960 did ball practice become part of training.'

Arsène Wenger is given much credit for the appliance and acceptance of science, enlightening his players and staff about such fundamentals as a good diet and football-specific physical training. But it was arguably in Italy that sports science was first integrated into the professional game and, more specifically, at the Milan Lab, which we'll touch upon in chapter 3. Founded in 2002, this innovative facility combined science, technology, IT, cybernetics and psychology for the very first time, while forging links with globally prestigious research centres like the Massachusetts Institute of Technology (MIT). The result? A 92 per cent reduction in injury rates among AC Milan players and longer careers.

Back on the domestic front, it can't be overstated how much the Premier League's 2012 Elite Player Performance Plan enhanced the understanding of sports science's role in developing both junior and senior players.

The Premier League's 2012 Elite Player Performance Plan enhanced the understanding of sports science's role in developing both junior and senior players.

'The EPPP transformed football,' explains Mo Gimpel, director of performance science at Southampton, regarded as one of the most forward-thinking clubs in football. 'Suddenly, things like GPS, measuring heart rate, taking saliva samples for hormonal profiles ... they were seen as important. Football very much had a closed mentality until then, and was certainly looked down upon by sports like rugby union that embraced innovation, both in training and nutrition. Suddenly, with the EPPP and the money in the game, there was an expectation from the FA that you protected players and their careers. We've gone from being dumb-asses to, in many regards, leading the way.'

Of course, for some people habits are ingrained and hard to break. Alan Pardew partly blamed his sacking from Crystal Palace at the end of 2016 on sports scientists and medical staff becoming too protective. 'When I started, the hard pre-season was a big thing, maybe too extreme among some,' Pardew told *The Times*. 'But I think we have come too far the other way – doctors under pressure at Premier League level, sports scientists protecting themselves a little bit, sometimes worried what the chairman will think if there is any injury.

'I'm not saying I don't value them [or] don't take their opinion. I just think we have become too protective. I wonder if footballers are at the level they can be physically. I think many can go further.' Pardew added that in his next job, the players would have 'a pre-season to remember'. That next job was at West Bromwich Albion, where a quartet of his most senior players had a mid-season bonding break to remember when they were arrested for stealing a Barcelona taxi. Pardew was sacked after four months.

Many footballing books focus on team tactics, the technical side of the game or the psychology of footballing success. Two books that detail the tactical and technical and stand the test of time are Jonathan Wilson's excellent *Inverting the Pyramid: The History of Football Tactics* and Johan Cruyff's autobiography, *My Turn*; while Damian Hughes, Professor of Organisational Psychology and Change at Manchester Metropolitan University, examines the mindset of arguably the greatest manager ever in his excellent book *How to Think Like Sir Alex Ferguson*, its themes relevant to everyone. There are, of course, also thousands of biographies, but few football books focus on the physical, and the ones that do are generally aimed at either academics or practitioners.

The aim of this book is to fill that hole by examining the application of sports science and cutting-edge technology to world-class footballing performance, all in an accessible narrative that will open your eyes to footballing science, not blind you with it. Of course, naturally there's a crossover with tactics – they heavily dictate physical training, as you'll see – and I specifically devote a chapter to the mindset of the modern footballer, as psychology and mental training is becoming an increasingly important component of training. I also examine the impact of technology on pitches, boots and balls, and how this has shaped the modern game.

But above all, this book focuses on the physical conditioning, training methods and equipment and nutritional strategies that have made today's game faster and more intense than ever before. Some of the tools are high-tech and arguably the preserve of professional footballers only; some are simply innovative ideas that can easily be integrated into your own footballing journey or that of your children, whether that's in the semi-professional or academy world, Sunday leagues, or just a six-a-side on a Monday night with Keith from Accounts. The focus is mainly on the men's game, as that is where most of the money and research is directed, but pretty much everything here can be applied to female players. And the women's game is growing so fast that I hope there will be more women's football-specific research to draw upon by the time of any second edition of this book.

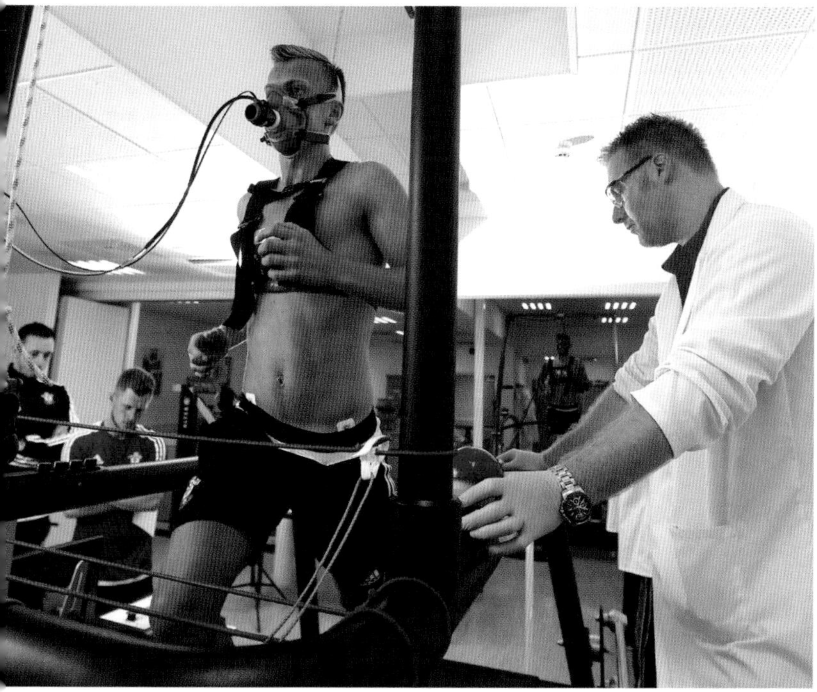

▼ Academics at Solent University test the fitness of Southampton's James Ward-Prowse

As you'll discover, I've been given privileged access to the world's best teams, from Manchester United and their neighbours City to Barcelona, Paris Saint-Germain, Ajax and many more. I've interviewed sports scientists, fatigue experts, chefs and nutritionists to understand what goes into creating the modern footballer. If you're sceptical, just remember what Sir Alex Ferguson said in 2013: 'Sports science is the biggest and most important change in my lifetime.' And he enjoyed a modicum of success. Sit back and enjoy the world of GPS-guided training, observe recovery via pitchside sleep pods and discover why Cristiano Ronaldo could never have mastered the knuckleball free-kick without ball technology.

THE NUMBERS GAME

Charles Reep is a man you've probably never heard of, but he arguably changed the football landscape forever. In 1950, Reep, an RAF accountant by trade and fan of Swindon Town, decided to bring a pen and paper (and a miner's helmet for evening kick-offs) to every single match he attended. With meticulous attention to detail, he would scribble down play-by-play diagrams of key moves. He'd eventually log and detail 2,200 games until the mid-1990s, spending up to 80 hours analysing a single match. In that time, Reep came to a realisation that would shape English football for years to come.

'Not more than three passes,' Reep opined during a BBC interview in 1993. 'If a team tries to play football and keeps it down to not more than three passes, they will have a much higher chance of winning matches. Passing for the sake of passing can be disastrous.' Charles Hughes, the Football Association's coaching director at the time, used Reep's findings to advocate the long-ball game that left England trailing behind their progressive international contemporaries.

While the conclusions that Reep drew from his data were questionable to say the least, he can rightly be called the father of football analytics.

GPS: Navigating the pitch

◄ GPS vests are now commonplace at professional football clubs. Here 'modelled' by Liverpool's Jordan Henderson (left) and Georginio Wijnaldum

How football's changed. There was a time, in the not too distant past, when your favourite team would be photographed in training with socks rolled down to their ankles, shirts billowing over too-short shorts and sleeves scrunched up to sharp elbows. Now, it's nearly impossible to find a professional footballer in training who is not wearing one of those tight black vests that resemble a crop top. These house arguably the most important – and certainly most visible – innovation to enter the world of football in the last 10 years: Global Positioning System (GPS) technology.

'GPS is pretty much mainstream across the world now,' says Canada's Darcy Norman, director of performance at AS Roma, who formerly worked with Germany at the 2014 World Cup. 'We used to rely primarily on heart rate to determine a player's intensity of training but now, with GPS, we can chart their sprints, distance run … a whole host of metrics that help refine training sessions.'

The two main players in performance-tracking are Australian outfit Catapult and Irish company STATSports. Both provide compression vests worn on the upper body into which a GPS sensor is slotted. As well as tracking a player's movements, the equipment also measures acceleration and picks up extra information on impacts, jumps and changes of direction.

GPS units use satellite-based technology, which operates by transferring signals between the unit located on the player and a network of satellites. This data transfer uses a sample frequency measured in hertz, and the higher the sampling frequency, the more accurately a player's position and movements can be tracked. 'When I was at Aston Villa, around six years ago, the sampling rate was around 1 hertz (Hz) per second,' says Matt Taberner, Everton's first-team sports scientist. 'Now we're up to 18Hz. Essentially that means the information's more accurate, especially as the players wear them in every training session.'

Although GPS trackers might be omnipresent in training, they are less common in matches. 'Despite the vests being close-fitting, players like a little more freedom

▼ Everton, pre-season training in Austria, use GPS units from STATSports

on match days,' says Taberner. 'That's why we use a match-data company who have a number of cameras in the stadium measuring movement at 25Hz. They pick up the coordinates of the players in terms of x, y and z, so you can cross-coordinate that with the information we have from STATSports for accurate results.' The cameras are able to track individual players to produce a bird's-eye animation of each player's movements during the match.

▲ GPS has measured Gareth Bale's top sprinting speed at 36.9km/h

Computer-based data capture has been around since the mid-1990s when a group of management consultants set up the sports analytics company Opta. This proved a ground-breaker but was – and still is – fundamentally a group of football fanatics scrutinising games while they unfolded, logging 1,600–2,000 notable events (passes, tackles, interceptions, headers ...) each match that were then instantly fed to Richard Keys and co. at Sky Sports, the media giant that gave Opta its first major contract as a way to beef up its football coverage. Opta were soon joined by rivals Prozone, a company that began life as a purveyor of the kind of massage armchairs that can be found in motorway service stations up and down the land. There are now further innovations like ProScout 7, a database that profiles over 130,000 players in more than 130 countries.

But it's the wearable GPS that has had the greatest impact on the football landscape. It tells us, for example, that Gareth Bale is currently the fastest player on the planet, registering a top speed of 36.9km/h, closely followed by Brazilian side Flamengo's lively winger Orlando Berrío at 36.0km/h.

Ever-increasing intensity

GPS research from the likes of Dr Paul Bradley, a sports scientist based at Liverpool John Moores University and consultant at Barcelona, is impacting how the game is played. Bradley is highly respected among his peers and in the football community in general, with over 60 papers published in the area of football science. But it was a piece of research published in November 2014 in the *International Journal of Sports Medicine* that grabbed the headlines...

Bradley and his fellow researchers examined how physical and technical attributes had changed between the Premier League seasons 2006–2007 and 2012–2013. And, in short, *everything* had changed – for the better. Across all positions distance covered increased, albeit by only 2 per cent, from 10,679m to 10,881m. That's negligible and, according to many experts, irrelevant, as distance run is not a game-changer.

More significantly, the distance of high-intensity running, defined as running at a speed between 19.8km/h and 25.1km/h, had increased by 30 per cent – from 890m to 1,151m a match. Bradley also showed that the number of sprints (over 25.1km/h) and total distance sprinted had shot up by 35 and 85 per cent, respectively. Bradley's work revealed that, with better pitches, more money at stake and greater accountability, the game was becoming more physical. Yes, technique and tactics still impacted on performance but both were diluted if the player who'd just executed a sublime stepover didn't possess the leg power to then accelerate away from the embarrassed defender.

The game had changed and players needed to change with it. 'So just make them work harder in training,' you might say. 'They're paid enough.' It's not that simple. Although money may enable footballers' lavish lifestyles, it has no influence over their muscle fibres. If muscles are pushed too hard, too often, like a stretched elastic band they'll eventually snap and lead to a spell on the sidelines. It's why the Holy Grail for fitness trainers and the football teams they work with is to marry peak performance with low injury rates. This is where the work of Australian sports scientist Tim Gabbett comes in. In the past, Gabbett has worked with Barcelona, Chelsea and Manchester City ... and all because of his development of a concept known as the *acute–chronic training load ratio*. Gabbett's model revolves around optimising training over a session and period of time so that the player grows fitter without breaking down.

Southampton's head of sports science, Alex Gross, says it's a model his club uses, and explains how the acute–chronic training load ratio works.

'Essentially, we might work in three- or four-week blocks,' he says. 'Chronic load is the average training load for the previous four weeks of training, while acute load is the specific load from that week.

'We want our chronic load, which is our fitness level, to be as high as possible. So we work back from when the players are under the greatest stress, which is generally the Christmas period where they could be playing three games in five days. If you get it right, you won't see big spikes in training load that can lead to injury.

'How I explain the acute–chronic model is that if you drink a bottle of wine every night and then go out at the weekend and have a bottle of wine and two pints, you'll probably be fine the day after. If you don't drink – so your chronic load is low – and go out and have a bottle of wine and two pints on a Saturday night, on a Sunday you'll be in a state. That's what the body is like. You really need to keep chronic load quite high to allow the players to manage periods of fixture congestion and help with recovery.' But don't take that as an invitation to pickle yourself in Pinot Noir.

The training load is calculated from a host of GPS-derived data like number of sprints and high-intensity runs. Heart rate will also be thrown into the mix. 'With that information, we might work in blocks of low week, medium week, high week, but each block is slightly increased – by around 10 per cent – so by December the players should be able to cope with those peak festive demands,' says Gross.

Through his research and on-field application, Gabbett discovered that an acute–chronic ratio of between 0.8 (acute load 80% of the chronic load) and 1.3 (acute load 130% of the chronic load) led to improved performance while cutting injury risk. Guided by that 'sweet spot' ratio range, coaches and sports scientists can then build on Gabbett's work to make things more specific and more relevant to the group of players in front of them. 'We've developed our own in-house tracking system, not an off-the-shelf one,' says Adam Brett, head of medical services at Brighton and Hove Albion, who formerly worked with England's national rugby team. 'With that, we've devised our own safe training-ratio range. It means we can be relatively confident that if we're in this "safe zone", the player will avoid soft-tissue injury and perform to his peak.'

That's no easy feat: overcook it and you risk injury; undercook it and you underperform. For Southampton things became more complicated when they qualified for the Europa League. 'In 2016–2017, because we were in Europe, we had to prepare players for three games in a week in a much shorter timeframe,' says Gross. 'So instead of a nice, steady 10 per cent increase, we needed a sharper increase. If there's a sharper increase in that acute load, the risk of injury is increased. You're pushing people potentially far too early.'

▼ Southampton employ the popular acute-chronic training model to maximise their players' fitness

Tactical periodisation

What complicates training prescription further is that – unlike cycling or running – football is a sport in which technique and tactics are just as important as physical fitness. You can't just have the players running up a hill, hitting a certain speed or distance, and expect to execute game-winning, tactically acute moves. That's where tactical periodisation comes in, a system of training that combines skills, fitness and tactical and mental awareness in each training session, often carried out at an intensity greater than during matches so that the players learn to react quicker, or be sharper, at crucial points in games.

'It's something we use and something I first came across when I worked with José [Mourinho],' says Celtic's head of performance, Jack Nayler, whose CV includes Chelsea and Real Madrid. 'It's also used by Carlo Ancelotti, Guus Hiddink … and similar models are used by Pep Guardiola and Mauricio Pochettino.'

This marrying of different components revolves around four key aspects of the game: transition from attack to defence; offensive organisation; transition from defence to attack; and defensive organisation. The genius of the manager and his team is to, as Mourinho said, 'Make operational our game model … the structure of the training session and what to do each day is not only related to tactical objectives but also to physical fitness…'

How this is applied on the training field is complicated. Like Nayler, Celtic manager Brendan Rodgers developed tactical periodisation under Mourinho when at Chelsea. Below is an example of how Nayler, Rodgers and the Celtic team train over seven days if Celtic are playing one game a week – rare in the 2016–2017 season when they played, on average, every 3.2 days before Christmas; then every 4.6 days after Christmas once they had been knocked out of Europe.

'We'd split a week without a midweek game into three blocks of two days: a recovery phase, a loading phase and a preparation phase,' says Nayler. Recovery takes place over the first two days after a game. The first day will involve low-intensity work at the gym, in the pool or on the exercise bike. Massage and foam-rolling are mandatory. The next recovery day sees the players back on the training pitch – again working at a low intensity but with football-specific elements and stretching throughout. 'We also do some small-space work with a high number of players so there's little room to run, meaning they're not really exerting themselves,' says Nayler.

'The loading phase is when we crank things up. It might involve small spaces, small numbers, and lots of deceleration and acceleration work. There'll be a high metabolic cost, so high heart and breathing rate but not really covering any distance.

'On the second loading day we'll open the space right up and allow the players to cover some high-intensity distance.

> Players need 36–48 hours to recover from any high-intensity running they do, so it can't be too close to the game.
>
> **JACK NAYLER** HEAD OF PERFORMANCE, CELTIC

▲ Celtic's Charly Musonda (right) is tackled by Zenit St Petersburg's Igor Smolnikov

We work with coaches to design drills that match the intensity. It's also a chance to do full-pitch tactical work where we'll hit high-intensity running speed and load. Our philosophy is that players need 36–48 hours to recover from any high-intensity running they do, so it can't be too close to the game. So we reach peak load three days out from the game to have further days to recover. These two "loading" sessions last for 75–90 minutes.'

And then it's the preparation phase, during which sessions are cut back to just over an hour at a time. Space is reduced on the first preparation day so players can focus on speed. 'They're not necessarily hitting top speed but everything about the session is quick, so maximum acceleration, speed of ball movement and speed of thought,' says Nayler. 'Most of this is ball work.

'Then a day before the game – the second preparation day – it's again a focus on the speed of ball and players' reactivity. It'll be competitive and short at around 45–60 minutes. Before that session, players will walk through tactics and set plays with the manager.' Then it's match day. But what if there's a midweek match? 'If we play Wednesday [after a Sunday],' Nayler adds, 'we'll still have two recovery days, so the day before the game will still be a second recovery day.'

These phases are also known as match day -1 (if game is one day away), match day -2, match day +1 … you get the picture. However it's termed, this micro-periodisation of a footballer's training plan can become complex and almost unmanageable when a team experiences success. 'Not being in Europe does have an impact, with less fatigue across the team,' says Nayler. 'It allows you to be more prepared for league games, mentally and physically … We've often debated among the staff whether the physical nature of the Premier League will prevent Premier League clubs challenging for league honours while challenging in Europe.'

Physicality dictated by tactics

Nayler has a point. Anecdotally, the English Premier League is regarded as the most physical in Europe. It's a theme picked up by Paris Saint-Germain's head of performance, Martin Buchheit. 'Maybe if you looked at average data over a year, there'd be a slight trend for the Premier League to be more demanding than the French league and other leagues, but that doesn't really tell you anything meaningful,' he explains. 'More important is understanding how different variables and game situations affect the physical output of the players. How much load we

▼ Variables like formation, tactics and the current score all impact on a player's physical output

▶ 'Studies show that when a team has a red card, the other players increase their running demands,' says PSG's Martin Buchheit

produce can vary by up to 15 per cent from game to game depending on the amount of possession we have or the pressing we do. In one game our midfielders might be happy to play around with possession and not need to run too much. Whereas another game might be all about high pressure or playing long balls and running. Tactics really play a big part...'

Cue further work by Dr Paul Bradley, who investigated the impact of team formation on high-intensity running and technical profiles in the Premier League. Based on analysis of 20 games, teams in 4-4-2, 4-3-3 and 4-5-1 formations enjoyed comparable amounts of possession; there were no significant differences in high-intensity running, either, though players in a 4-5-1 formation performed less high-intensity running when their team had the ball but more when they didn't. And you've got to feel for the strikers in a 4-3-3, who got through 30 per cent more high-intensity running than their counterparts in the other two systems. Also, pass completion was higher in the 4-4-2.

'Studies show that when a team has a red card, the other players increase their running demands enough to make up for the missing player,' adds Buchheit. 'The scoreline has a huge impact, too. As soon as there is more than two goals difference between the teams, the winning team run less because they think the game is over. If you're losing 2–0, your run data goes up as long as there's reasonable chance to get back in it. If it's 3–0 or 4–0 with 20 minutes to go, both teams' run data goes down.'

Buchheit's point about the scoreline might help explain why teams like Barcelona and Real Madrid consistently outperform English teams in Europe. En route to destroying Juventus 4–1 in the 2016 Champions League Final, Madrid won four of their final six La Liga games of the season by three goals or more – enabling them to save energy for that Cardiff final on the horizon. Fast-forward to October 2017 and Crystal Palace, bottom of the Premier League with no points and no goals scored, are beating champions Chelsea 2–1. Strong competition equals greater physical demands during the whole match equals reduced chances of winning the Champions League.

The Catalan innovators

Clearly this is quite simplistic and overlooks the attention to detail progressive clubs like Barcelona pay to data and training. In March 2017, Barcelona launched their Innovation Hub, which 'aims to send the club to the forefront of the sports industry as a centre for knowledge, innovation and technology'. The club plan to work with universities, research centres, start-ups, sponsors and entrepreneurs to lead the way in five key areas: team sports; sports performance; medical services and nutrition; technology; and social sciences.

While espousing the collective benefits of knowledge sharing, rather cannily Barcelona have brought the world's foremost experts to them by virtue of an annual conference, the Sports Technology Symposium, which attracts leading lights from all sports, including rugby's All Blacks, the NBA's LA Clippers and Ferrari in motor racing.

In May 2017 and June 2018 Barcelona also hosted the Future of Football Medicine conference, an annual event organised by the Isokinetic Medical Group, an Italian sports injury consultancy recognised as a FIFA Medical Centre of Excellence. The three-day conference focuses mainly on the latest developments in injury prevention and rehab. However, it also touches upon peak performance and managing training loads and, in 2017, representatives from Barcelona FC hosted their own two-hour symposium.

One of the most intriguing and insightful presentations showed how the Barcelona staff varied the intensity of their training drills depending on where they were in the week. The heart of their training model revolved around 10 levels of specificity (see box), with each incorporating at least two factors from the technical, tactical and physical.

The coaching staff can then play around with where each level and session fits into the weekly plan in order to achieve the right kind of training load for each day. For example, match day -3 is when the staff roll out the training session with the most running. According to the club's own data, Barcelona's players run an average of 104.7m per minute during an actual match. So rondos, for example, would not provide enough intensity as they involve a running rate of only 38.5m/min (37% of match day output). The coaches would be more likely to choose situation games (98.2m/min, 94%), big-pitch positional games (95.4m/min, 91%) or even technical circuits (90.1m/min, 86%).

The interplay of these levels, which can be slotted in and out depending on the physical demands of the week and the head coach's tactical desires, is impressive. The symposium also made it clear that Barcelona, like many clubs, focus on quality over quantity.

Barcelona's training levels

Barcelona's 10 levels of specificity are as follows:

1 Conditioning circuits, which could be gym-based
2 Technical circuits on the training pitch
3 Direct circuits
4 Rondos (players in the middle of a player-formed circle chase the ball)
5 Small-pitch positional games (players occupy positions similar to those in a regular game – e.g. centre-back, central midfield – but on a smaller pitch)
6 Big-pitch positional games (players occupy positions similar to those in a regular game and play on a normal-sized pitch)
7 Small-sided games
8 Situation games (i)
9 Situation games (ii)
10 Match day

Keep it short and intense

'Speed, particularly repeatable speed, is a strong differentiator between Premier League players and Championship footballers,' says Southampton's head of sports science, Alex Gross. 'Total volume and distance is actually similar but it's the top end that's different. And that would be across each role. That's why when you watch the Premier League, it looks faster – every action is just that bit quicker than you would see elsewhere. Running – and reacting – faster than the other team has more impact than simply running further. In fact, if you look at our under-18s, they actually cover more distance than our first team will in a game. But they're not technically quite as proficient.'

Training that repeatable speed is one reason why many teams raise the intensity of sessions and reduce the volume. As well as football-specific performance improvements, players also benefit from what Tim Gabbett calls the *training-injury prevention paradox*: namely, the harder you train – within reason, of course – the lower your chances of injury. It's something we'll cover in more detail in the next chapter.

Arsenal's Australian director of high performance, Darren Burgess, is a fan. Burgess spent three years at Liverpool as head of fitness and performance under Kenny Dalglish, Roy Hodgson and Brendan Rodgers. Before his recent switch to Arsenal, he'd been working for Australian rules football side Port Adelaide. He also worked with the Australian national football team during the run-up to the 2010 World Cup in South Africa.

'One of the things I've learned in football and Australian rules football is that players must work hard,' Burgess said at the 2017 Future of Football Medicine

▶ Like many clubs, Barcelona use rondos as both a training and warm-up drill

▲ The diligently defensive Carles Puyol (left) required different physical attributes to the relentlessly attacking Dani Alves (right)

conference. 'That's why we measure the worst-case scenario for every player. We have some who have to run hard for 190m/min for five minutes; others 130m/min. We make sure that every week the player is exposed to maximum velocity so that his muscles – especially the hamstrings – adapt to this speed.'

After spending hours analysing his data, Burgess concluded that his teams executed, on average, 20 per cent more sprint work in training before games that they won compared to games that they lost. 'It's something I noticed at my time in Liverpool, too, albeit after I left!' he said. 'I reflected on over 100 games and it was the same – when we trained hard, we enjoyed a greater win percentage. In my experience, the more aggressive approach gives you a better outcome.'

Positional demands

This comes back to maintaining a high level of fitness (from chronic load) and managing spikes (acute load). Burgess also touched upon different players having a different 'worst-case scenario', which often boils down to their position on the pitch. A 2015 study led by Barry Drust, Professor of Applied Exercise Physiology at Liverpool John Moores University, who has worked for Liverpool and England, examined the physical demands by position of the outfield players of Rosenborg, who have won the Norwegian First Division a record 25 times. Drust collected the players' GPS data over 15 games. Not surprisingly, he found that wide midfielders ran the most on average, at 12,320m. Central defenders ran the least (10,219m). The wide midfielders also sprinted the most (1,168m), whereas the full-backs sprinted the least (491m). (Rosenborg's full-backs were very much in the defensive vein.)

It's an area Jens Bangsbo, Professor of Exercise Physiology at the University of Copenhagen, knows well. Bangsbo also works with the Danish national football team and has done so since 2004. Before that, he was assistant coach at Juventus between 2001 and 2004. He's that rare beast in football who holds a high-ranking academic role in sports science as well as a UEFA Pro Licence coaching qualification.

Bangsbo is a big fan of individualised training. He focuses not just on position but also what the manager asks of the player. 'Though position clearly influences a player's physical output, I'd say the tactics employed by the coach are even more important,' he says, 'but that's not limited to rigid structures like 4-4-2 and 4-3-4. Take Dani Alves and Carles Puyol when they both played for Barcelona. While Puyol would often play in the centre, at times he also played full-back. So they would be the two full-backs in the same team. But while Puyol might be running at a high intensity for 1,000m, Alves would easily run double that. I've also seen players in the same positions where one has completed 44 short sprints and the other 11 longer ones.'

Alves was, and still is, a unique attacking threat, known for his late sprints in behind the opposition defence. Countless times he'd meet a pass from the centre-left midfield position, usually from Xavi or Andrés Iniesta, and either score (14 goals in his 247 appearances for Barcelona) or, more commonly, slide the ball sideways for Lionel Messi, Luis Suárez or Neymar. In fact, as the Brazilian's Barça career drew to a close in May 2016, he became just the third player in La Liga history to register 100 assists, after Messi and Luis Fígo.

Puyol, on the other hand, who retired in 2014 after 392 games for Barcelona, was a defender in the traditional sense – combative, stoical and intent on keeping a clean sheet. So much so the Camp Nou fans nicknamed him 'The Wall'.

'It begs the question: how should you train these two guys?' asks Bangsbo. 'Differently? Of course. But, historically, individualisation of training has been neglected. You'd simply have a fitness coach standing ready with a whistle while the players line up to sprint. One blast and they're off, sprinting exactly the same distance, despite this sort of data from GPS telling us that's not match specific.' (As an aside, Bangsbo also suggests that sprints in training should start with a jog rather than from a standing position, again to better replicate the match situation.)

Clubs are listening, with Southampton's Alex Gross saying they'll often pull players out of a certain drill if the training GPS metrics don't match what's needed for their role on the pitch. 'That might mean more sprints, which they're not always overly happy about,' Gross says.

Tim Gabbett is a proponent of grouping. As it sounds, this is where you split the team into three training groups: the high-loaders, who are the players who can handle heavy loads and spikes in intensity; the in-betweeners, who are neither robust nor fragile; and the fragile players. No one likes to be called 'fragile' but football supporters are fully aware of players who seemingly spend more time with the physio than their teammates. Take the unfortunate Daniel Sturridge, who has missed more than 100 games through illness or injury since signing for Liverpool in 2013.

Transfer awareness

You can see how the intelligent, progressive manager builds up an understanding of the physical demands of the tactics he wants to employ. This clearly has implications for transfer policy. 'With any new signing, there are a number of factors you have to be aware of as a department. One is your team's style of play and tactical set-up,' says fitness trainer Nick Grantham, who works with West Bromwich Albion. 'At West Brom there's a certain make-up of what our players look like and what's expected of them.

'Now, if you're bringing in a player who's scoring for fun in an overseas league where perhaps the intensity isn't as high – the Russian league, for example – you have to be mindful about managing their load and letting them get used to the pace of the Premier League. It can take up to a season for players to bed in. You might also sign a player from a more successful team where they served as an impact player, coming off the bench. But you want them as a main player, so they're now sprinting hard for 90 minutes instead of 20. That can be a gamble.'

Positional- and tactical-specific statistics were behind Manchester City manager Pep Guardiola's decision to spend £118 million on a trio of full-backs in a two-week period in the summer of 2017: Benjamin Mendy (Monaco, £49 million), Kyle Walker (Tottenham Hotspur, £43 million) and Danilo (Real Madrid, £26 million).

▼ Pep Guardiola spent £43 million on Kyle Walker (right) as his tactics relied on a powerful, fast full-back

The modern full-back is asked to perform a near-impossible task: press high up the field when the opposition have the ball in their own half; contribute to attacks when the ball is won; and recover to perform defensive duties when the press is broken. Back at Bayern Munich, Guardiola used Philipp Lahm and David Alaba in a role that became known as the 'inverted full-back'. When defending, the duo would

APPLY THE SCIENCE

Train to the beat

Measuring heart rate is still key to reaching optimum fitness for players of all levels

Football is a game of 90 minutes plus added time (significant if you ever faced Alex Ferguson!). Physically, it comprises periods of walking, jogging, sprinting and jumping, some of which can send your heart rate through the roof while others have it nestling at an easy-to-talk level. Professional football clubs use GPS data in tandem with heart rate measurement to monitor heart rate in relation to metrics like sprinting speed and acceleration rate.

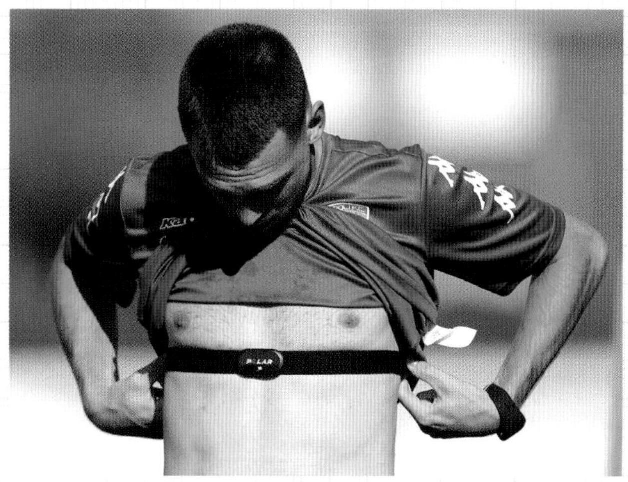

▲ Heart rate monitors, like this model from Polar, help to train different parameters of football-specific fitness

Recreational players can use a heart rate monitor like those from Garmin and Polar to improve their game without needing to invest in a costly GPS system. But how do you get the best from a heart rate monitor? First, you need to discover your maximum heart rate (maxHR). A simple way to do this is to run 1,000m as hard as you can and then note your heart rate at the end of the run. This will be your maxHR.

With that information, you can now set about improving fitness. 'To get fitter, your heart must be working close to its maximum,' according to Southampton's director of performance science, Mo Gimpel. 'When you're training, you're looking for a stimulus to make you fitter. This means working at around 80–85 per cent of your maximum heart rate.'

That's why interval training is so common – you can train at a high intensity to stimulate football-specific adaptations, like long sprints, and then rest and repeat as you would in a game. But it's also beneficial to throw in some long, slow runs to establish a solid physiological foundation from which to build speed.

Examples of heart-rate-based sessions from Polar include:

- Long, slow running at 70 per cent HRmax for 45 minutes
- Lactate-threshold running for around 20 minutes at 85 per cent HRmax
- 15/15 training to build speed endurance, namely 15 seconds at 90–95 per cent HRmax followed by 15 seconds at 70 per cent HRmax, repeated for as many times as you can tolerate
- 30-second efforts where you all-out sprint for 30 seconds followed by four-minute recoveries until HR reaches 60 per cent HRmax – this builds top-end speed.

act as traditional full-backs, but when Bayern had the ball, they'd slot into the centre and provide Guardiola with two extra central midfielders. The midfielders could then play higher up the field, as could the attackers, meaning Bayern would dominate possession and overwhelm the opposition in the final third.

At City Guardiola has reverted to more powerful players down the touchline. The contribution made by Walker and Danilo to the club's 2017–2018 title win suggests he chose well. (Mendy was out injured for most of the season, but has come back strongly.) Critics questioned the money spent, but Guardiola knew full well the importance of this type of player to his strategy. They are looked upon with envy by opposition managers.

'We would love to use our wing-backs more and play more like Barcelona, Man City and Tottenham Hotspur, but physically that's a stretch for our players,' said Dave Tenney when he was head of sports science at Major League Soccer outfit Seattle Sounders (he has since moved to a similar role at basketball team Orlando Magic).

The GPS innovator

Tenney has carved a global reputation for maximising his resources through innovation, specifically his use of GPS. 'Our understanding of the data and what to look for is changing,' Tenney says from his US base. 'When we started, the metrics we focused on included distance covered and roughly indicated that if you ran more than the other team and had more shots, you'd win. But we discovered that wasn't very accurate.

'Our understanding has evolved. We now know that the metrics that have the greatest impact are: your ability to close down the opposition defensively at high velocity over the course of a game; your ability to dominate space and your opponent; and, when you have the ball, your ability to bypass your opponents. They call that "packing" in Germany.'

While interested in specific players' actions, Tenney is focusing increasingly on 'interactions': 'Instead of collecting data solely on Xavi, for example, you'd see how he interacted with Iniesta. You can also measure, say, the distance between the back four and the holding midfielders. Will it become stretched as a game goes on or stay the same?

'With our optical technology we noticed that in the first 25 minutes of a match we might keep to our game plan, which is looking to build from the back,' Tenney continues. 'Then somewhere around the 25-minute mark we start to tire a little and you see things fall away. Come the 70th minute, as players fatigue and the game becomes more frantic, interaction between players drops off. Perhaps the left-back, who'd been following the game plan, has pushed too much. And the centre-back isn't talking to him as much as he was earlier on. Interaction's not quite as good as before.

'Our striker should then fall back and make us really compact, but he's tired, too. So the holding midfielder begins to drop deeper into the back four and the distance between holding midfielder and striker grows way too big, meaning the other team can play right through there. Now, is that down to discipline? Or fitness? Or tactics? It's probably the interaction of all those factors, and really getting a handle on that interaction to influence training sessions and selection is the way ahead.'

Tenney is not the only one focusing on GPS version 2.0. Head over to Holland and the training grounds of Ajax, Vitesse Arnhem and Feyenoord resemble GCHQ, with pieces of radio-communication equipment matching the number of balls, cones and talented teenagers.

'It's a system called Inmotio,' explains Ajax's head of sports science, Vosse de Boode. 'GPS is great as long as you talk in terms of height, total distance, intensity running ... but as soon as you go into accelerations, cuts and turns and position on the field, it isn't quite as accurate and detailed. That's why we use Inmotio – we feel we can define football positions more accurately.'

▼ AFC Ajax's head of sports science, Vosse de Boode, monitors data like players' accelerations

That added detail comes from a dome camera that monitors the players, while also sending out a radio frequency. As in the other GPS systems, each player wears a vest containing a transponder, which the antenna recognises to collect a huge amount of data – up to 1,000 pieces per second – but there's one key area that de Boode identifies as its main selling point.

'Accurately monitoring accelerations is huge,' she says, highlighting that while other GPS products might measure this metric they're nowhere near as detailed as the system developed in 2012 by Austrian technology firm Abatec AG in partnership with Dutch sports scientists. 'For example, a small-sided game might not contain a lot of high-intensity running, but it might be super intense when it comes to cuts and turns and more technical actions. If we can't accurately measure accelerations and decelerations, we miss a big part of the game of football and training.'

Expected goals

OPTA'S XG TOOL HELPS CLUBS PLAN FOR MORE CONSISTENT RESULTS

'He had no right to score from there.' 'I could have put that one away.' 'He should have had a hat-trick.' Three classics from the commentator's and fan's handbook, but how true are they? Well, now empirical evidence can replace subjective frustration thanks to Opta's expected goals (xG) tool, which appeared on BBC's *Match of the Day* for the first time at the start of the 2017–2018 season.

The model gives each chance in a game an expected goal value (EGV), which is a calculation of the likelihood of a goal being scored from that chance, based on Opta's analysis of more than 300,000 shots from previous matches. EGV is based on a number of factors, such as where the shot was taken from, proximity of defenders and the nature of the attack (e.g. a counter-attack, direct free-kick or penalty).

From an analysis of every shot's EGV in a match, an overall 'expected goals' (xG) figure can be placed on each team. The xG is the number of goals that the team would have been expected to score in that

match. If a team has a higher xG figure than actual goals scored, this may be down to wasteful finishing or good goalkeeping, or both. Conversely, if a team is scoring more goals than its xG then this could be due to moments of individual brilliance from an attacker or errors from the opponent's goalkeeper.

Why is this useful? Because it gives an indication of whether a team's results are based on sustainable factors like the consistent creation or denial of chances, or less reliable factors like sensational/woeful goalkeeping or good/bad luck. The xG figure evaluates the quality of a team's chances, whereas 'shots on goal' does not discriminate between a 35-yard rasper and a missed open goal from close range.

Here are the actual results and the xG results of a sample of Premier League matches played over the weekend of 14–15 April 2018:

FIXTURE	ACTUAL RESULT	XG RESULT
Southampton v Chelsea	2-3	1.29-0.66
Liverpool v Bournemouth	3-0	2.64-0.54
Tottenham Hotspur v Manchester City	1-3	0.45-3.62

You can see that the games at Liverpool and Spurs pretty much went to plan, whereas Chelsea carved out what would appear to be a 'freakish result', having scored three goals when their xG figure was less than one. For Chelsea this match harked back to their title-winning season of 2016–2017, during which their attack scored 22 more goals than they were expected to based on their overall xG figure for the season.

This was far and away the highest figure in the league that season; Tottenham were the next closest team, with 15 more goals scored than expected. Manchester United actually scored five goals *fewer* than expected, giving Chelsea a 27-goal advantage in this respect over their traditional rivals.

Sadly for Chelsea, during the 2017–2018 season the result at Southampton was the exception not the rule. With manager Antonio Conte deciding to discard Diego Costa, their top scorer, before the season started, Chelsea's cutting edge blunted considerably. Their xG advantage was greatly reduced – they scored just three more goals than predicted and they finished fifth, missing out on a Champions League place by five points. Their goals tally of 62 was 23 down from the previous season.

The impact a key striker can have is highlighted when you look at xG by individual player. Take Harry Kane's 2016–2017 season. The Spurs striker's total chances (not including penalties) were worth 13.86xG, but he actually scored 24 goals from them, meaning he converted 10 more shots than would be expected of the average player. Meanwhile, Sergio Agüero's total non-penalty chances were expected to yield 16.58 goals, compared to the 16 actual goals he scored from them. This means the Manchester City forward was performing at almost exactly the level Opta would expect him to.

◀ According to xG, Southampton should have won a game they lost to Chelsea 3–2

Hitting peak acceleration

One of the leading advocates of the Inmotio system is Raymond Verheijen, a Dutch coach known for his innovative periodisation ideas.

Verheijen worked with South Korea at the 2010 World Cup, where he monitored each and every session with Inmotio before and during the tournament. What he discovered had a significant impact on how he trained the players. When the squad assembled for a national team camp in mid-May before heading to South Africa later that month, they undertook a test comprising eight sprints with minimum rest. Repeated sprints are a key component of elite footballing performance, so this was a good benchmarking activity.

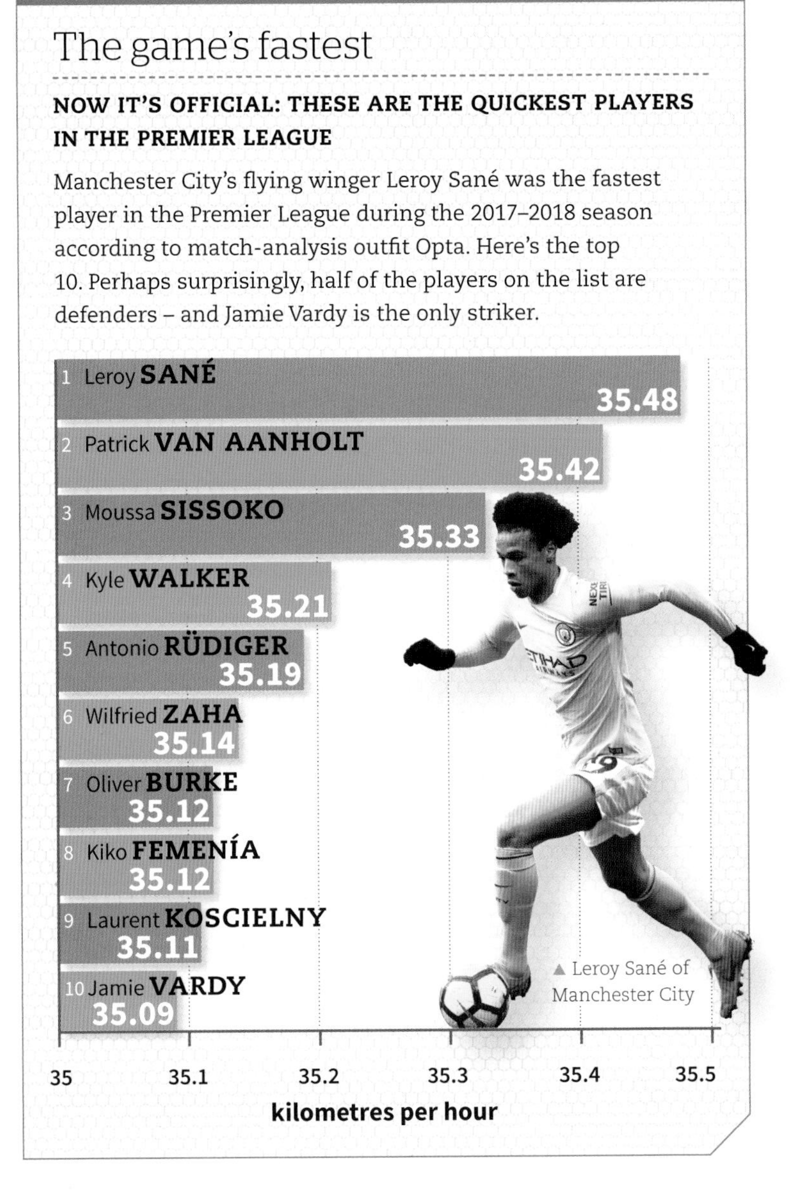

▲ Leroy Sané of Manchester City

Verheijen observed that the average peak sprint speed measured 7.1m/sec on the first sprint, dropping to 6.7m/sec on the final effort, a 6 per cent change. 'This insignificant decline in peak speed was proof that the factor "maximum speed" in football does not make the difference [when it comes to game-changing situations],' commented Verheijen.

However, he also noticed that peak acceleration plummeted from 6.6m/sec2 in sprint one to 5.1m/sec2 in sprint eight – a 23 per cent drop. This was a more significant finding for Verheijen, because 'a player who in the last quarter of an hour still wants to get past his direct opponent needs to maintain maximum explosive actions'.

Verheijen set out a training plan to maintain players' peak

acceleration, including small-sided games of four against four and three against three, plus variations like four blocks of four-minute matches with two minutes rest between. This would force the players into more actions – and, crucially, more *explosive* actions – to find space, find the pass and execute the move efficiently.

When the test was repeated in South Africa three weeks later, peak acceleration decreased from sprints one to eight by just 14 per cent, despite training taking place at altitude. The South Koreans took this improved football-specific fitness into the tournament with the aim of reaching the knockout stages, a feat they achieved by finishing second in their group behind Argentina before losing 2-1 to Uruguay in the next round. 'This proved that it is not so much the speed of the football player [that's important], but how quickly a player achieves this speed,' Verheijen concluded.

It's certainly an attribute players like Paul Scholes, Frank Lampard and Michael Owen maximised in their careers, their ability to rapidly reach top speed over a short distance often allowing them to find space to harm the opposition.

In-game decisions

Some of the GPS products, such as Inmotio and STATSports' new Apex system, now provide real-time data so changes can be made in the present rather than the future. Ajax's Vosse de Boode says, 'If we see a certain pattern that's game-specific but at a lower intensity than it should be, we immediately tell the trainer so they can alter things. This really helps our communication with the coaches. They don't want to hear after the match that we should have done something differently.'

FIFA has allowed players to wear GPS trackers during matches since 2015, but only in March 2018 did it permit teams to use real-time data to make decisions during the match itself, while the Premier League allowed the use of electronic devices on the bench from the 2018–2019 season. So a coach will now be able to identify, for example, the players whose acceleration rate is dropping off the most and use this to alter the formation or plan substitutions. This has been standard practice in Australian rules football for a number of years. Although players have found the tight-fitting vests restrictive during matches, some teams do wear them – notably the Brazil team during its 2018 World Cup matches in Russia.

Cynics could argue that the rise of GPS use and data analysis arose from the money pouring into the game. If clubs are going to pay players like Alexis Sánchez £300,000 a week, they want them held to account. But that ignores the true impact GPS is having on training professional footballers to perform at their optimum, for their specific position within a given tactical framework, in modern-day football. The human body is complex. So is sport. You cannot train a professional footballer in the same way as an endurance cyclist. Data and measuring tools might lack the aesthetic appeal of a Ronaldo stepover or a Cruyff turn but they cater for this physiological difference, transforming the pace and flow of the game. And, as we'll discover, GPS is just the tip of the technological iceberg…

SAP
27

FAST, FIT AND FLYING

It's the early 1990s and German club TSG 1899 Hoffenheim are playing to sparse crowds – again – in Sinsheim, a municipality of Baden-Württemberg. The obscure amateur side battle a quagmire of a pitch and part-time opposition in the eighth tier of German football. The club steadily improve and by 1996 are competing in the German sixth division. Better, but still a tiny dot compared to the sharp-shooting, multi-millionaire behemoths at Bayern Munich and Borussia Dortmund.

It's now 4 April 2017 and Bayern, brimming with stars including Arjen Robben and Arturo Vidal, head to the Rhein-Neckar-Arena, a 30,150-capacity, €100 million stadium opened in 2009 on the edge of Sinsheim, a town whose population is only 36,000. The Bundesliga leaders, who've lost just once all season and would go on to win a record 26th title, line up against Hoffenheim, who are a history-defining third in the league.

Under the watchful eye of their 29-year-old coach, Julian Nagelsmann, Hoffenheim beat the five-time European Cup winners 1-0 thanks to a sublime Andrej Kramaric half-volley. It's the first time they've beaten Munich in the Bundesliga and it all but secures their qualification to the European Champions League for season 2017–2018. (Hoffenheim put in a good account of themselves, but would lose to Liverpool in the Champions League group stages. Liverpool, of course, would go on to make the final, losing 3–1 to Real Madrid.)

◀ Appliance of science: in 2017, Hoffenheim's Andrej Kramaric scored the winner against the Bayern giants

The remarkable rise of TSG 1899 Hoffenheim gained momentum in 2000 when Dietmar Hopp, a billionaire software entrepreneur, began financing the club he had played for in the youth team. Major investment in players and that new stadium saw Hoffenheim win promotion to Germany's top flight for the first time in 2008. They have been there ever since.

Hoffenheim's technical heart

Hoffenheim's transformation has been remarkable; less remarkable perhaps is that a club funded by a computer specialist has maximised technology in its search for marginal gains. But you wouldn't know by the look of the club's training base – a converted castle in Zuzenhausen. In the 1970s and 1980s it had been a nightclub where bands such as the Sex Pistols and the Stranglers played. Incongruous as it may seem, this is the home of the technological revolution that's seen by many as key to Hoffenheim's rise through the ranks.

Inside, it resembles something out of an Arthur C. Clarke novel and features a plethora of cutting-edge training tools. There's the Footbonaut, first brought to the world's attention by Borussia Dortmund at the end of 2012, where a machine pumps balls out to the player, who has to control and pass to a randomly selected square on one of the four surrounding walls. Sharpened ball skills are the aim.

Then there's the Helix, a cognitive training tool featuring a giant 270° screen. In what resembles an oversized PC football game, animated players from two teams appear in front of the real player on the wraparound projection screen. After a momentary stationary phase, the virtual players dart around the pitch shown on the screen. Once they stop, the human player has to identify the players who were running around. Stadium sounds and volume levels can be altered to increase or decrease stress levels, the aim being to improve peripheral vision and the ability to track multiple objects simultaneously.

▼ Hoffenheim's Helix aims to improve the players' peripheral vision

Head outside and the technology continues. Ready for the 2017–2018 season, Hoffenheim installed the Videowall. The 6 x 3-metre screen is positioned halfway along the sideline of their main training pitch. Four cameras – two on a scaffolding tower beside the screen and one camera behind each goal – send a live feed to the screen. Nothing particularly innovative there, you might think, but these cameras are controlled via iPad by the training staff, who can stop, rewind or fast-forward to particular points of interest, giving the club's forward-thinking head coach, Julian Nagelsmann, a chance to explain specific situations in vivid detail to players while they are still on the training pitch.

Like the Helix, Hoffenheim's SpeedCourt improves cognitive ability, but it is also designed to increase reaction time and speed of movement. The player stands on a square in the centre of a court measuring 7m by 7m. Surrounding the player are eight smaller squares: one to the front and one behind; one on each side; and one in each corner. On the wall ahead is a TV screen, which reproduces the court layout.

When one of the squares lights up on the screen, the player sprints sideways, diagonally, backwards or forwards to the corresponding square on the court, then sprints back to the centre before heading to the next square that lights up. The intervals gradually shorten to stimulate quick feet. It's similar to the dancing squares you might find in a seaside amusement arcade but with one big difference: the floor is packed with sensors to measure aspects like the amount of force through each leg, and ground contact time, which is the length of time the foot is in contact with the ground during each step. (Also, there's nowhere to buy candyfloss.)

The developers of the SpeedCourt claim that it cuts ground contact time by 30 per cent. Why is this important? The theory goes that reducing the length of time the foot is in contact with the ground reduces the amount of force driving through each leg. Less force means less load through the limbs and, it is claimed by the developers, a reduction in injury. It also leads to a faster cadence (number of strides per minute). The result? Theoretically, healthier and faster players.

It's a logical conclusion, though independent literature demonstrating that shorter ground contact time is associated with fewer injuries is thin on the ground, if not non-existent. In fairness, it's notoriously difficult to pin down the cause of an injury unless it happened as the result of a specific incident. And in the case of ground contact time, the argument seems persuasive enough even if not supported by research in the labs. There's certainly more visible evidence behind those cadence claims. Away from football, everyday runners tend to average between 160 and 170 strides per minute; elite runners tip over 200.

Hoffenheim aren't the only ones who are convinced of the SpeedCourt's merits: Bayern Munich, Bayer 04 Leverkusen and Real Madrid also use it. Perhaps surprisingly, no Premier League clubs have invested in the system. The only one currently to be found in England is at the Running School in west London, established by Mike Antoniades, a specialist in run movement re-patterning and rehabilitation after injury.

Working the nervous system

The SpeedCourt has become an invaluable tool for Antoniades in what he believes is a neglected area of performance at many clubs. 'In European leagues, like Italy and especially Germany, the teams work a lot on speed,' he says. 'Apart from at the bigger clubs like Manchester United and City, that rarely happens in England.'

It's partly down to the congested fixture list, says Antoniades, but also, he claims more cuttingly, to a lack of understanding. 'A lot of people in football and fitness feel that speed comes from strength – it doesn't. Strength will protect you from injuries; speed comes from how quickly the nervous system fires the motor units. So what you need to do is work on specific exercises that will stimulate the nervous system.'

Antoniades has devised a series of workshops and assessments to increase a player's speed, and has worked with Hoffenheim – he was heading over to Germany the day after our interview – as well as Bayern Munich. Specific speed drills remain Antoniades' intellectual property, but 'there's some quick-feet stuff on our website'.

According to Antoniades, the key is biomechanics, analysis of the forces acting on the body to optimise muscle performance and durability, 'because football is a sport that's multi-directional and involves contact. The statistics that they show on television about this player covering 11,000–12,000m, that's totally irrelevant. What's relevant is how many times they've touched the ball; how many turns on the left-hand side; how many on the right; how many short sprints in that position in 90 minutes at full speed at intensity. And that's all improved by working on your biomechanics'.

> It's not about how a player looks but whether they are hitting the numbers. Do they have explosiveness and quickness of feet?
>
> **MIKE ANTONIADES**
> RUNNING SCHOOL

Pre-season is the key period for Antoniades and sports science staff in general, as it's when players' fitness is tested in order to set benchmarks against which they will be re-measured at a later date. He'll assess players both on the field of play and in the laboratory, with and without the ball, videoing them to later analyse football-specific movements like single leg push-offs, accelerations and decelerations, and jumping. 'It takes around five to six hours to analyse the whole team and this process will be repeated in the middle of pre-season and the end of pre-season when the players should be at their fittest,' Antoniades explains.

After that first pre-season test, each player is given their biomechanical passport, including video clips of themselves sprinting. Each will have individual areas to work on. And it's this bespoke advice that is key as, when it comes to a footballer's running technique, there's no one-size-fits-all solution. 'Take Yaya Touré. When he runs he looks quite cumbersome. But he's pretty fast. It's not about how a player looks but whether they are hitting the numbers. Do they have explosiveness and quickness of feet?'

Positional analysis

Antoniades also accounts for the demands of different football positions. Take a centre-back. One of their major roles is to defend high balls from set plays, so they need to have good jumping ability. Antoniades pays particular attention to their ground contact time when leaping. For example, if he notices that one leg has more ground contact time than the other when a player is jumping with both feet, he'll see this as a warning sign of imminent injury. 'We'd then look to correct that with biomechanical work,' he says. Antoniades argues that this structural focus and use of the SpeedCourt is why Bayer Leverkusen, another of the teams he works with, has one of the lowest injury rates in European football.

Pre-season is the ideal – possibly only – time that clubs can integrate this speedwork into their existing programme, because it takes about six weeks and a lot of repetition to change the neural map so that a new technique becomes an automatic process. As it happens, six weeks is the typical duration of a Premier League club's pre-season.

As for how each biomechanical speed session breaks down, you're looking at 30 minutes or more each day for the first seven to 10 days, and then three times a week to reinforce those new pathways. 'The more you run in a certain way, the greater the number of neurons that attach to the muscle and then they multiply,' explains Antoniades. 'The stronger this connection is, the quicker the messaging goes to the brain, the more muscles get recruited, the faster the movement.'

As well as recruiting more neurons, key to players increasing speed is a white, pearlescent substance known as myelin. This sausage-shaped layer of dense fat wraps around the nerve fibres in much the same way that insulation wraps around a wire. It maintains a strong electrical signal by stopping electrical impulses leaking out. The thicker the myelin sheath, the quicker the signal travels along the nerve fibres. And you achieve that by practice.

'As you repeat any movement, the myelin becomes thicker,' explains Antoniades. 'If you look at violinists, the myelin in the hand/arm holding the violin will be a lot thicker than in the hand/arm holding the bow because the violin hand is working through more complex processes. The same applies to football – players who have done more speed work will, in general, have a thicker covering of myelin over the nerve fibres involved in the movement.'

The advantages of this neurological transformation are clear. But what about when the season kicks in and time is tighter, because players are competing twice a week and travelling all over Europe? That's not a problem, says Antoniades, as you can maintain this new technique throughout the season with just two 15-minute sessions a week.

That's the ideal. But the unpredictable nature of football means Antoniades is often called up at late notice to work his biomechanical magic. 'Before the 2016 European Championships I worked solely with one of a national team's centre-halves

Neurological benefits of playing FIFA

IT'S NOT JUST EFOOTBALLERS WHO ARE BENEFITING FROM GAMING

'Andrea Pirlo has proclaimed that "after the wheel, the PlayStation is the best invention of all time",' wrote football journalist Rory Smith in a 2016 piece for the *New York Times*. 'Zlatan Ibrahimovic wrote in his autobiography that he "could go 10 hours at a stretch" playing soccer video games early in his career. John Terry used to host Pro Evolution Soccer get-togethers for his Chelsea teammates on the eve of each season.'

Smith also focused on the exploits of Arsenal midfielder Alex Iwobi, who'd watch the tricks of Ronaldinho on FIFA and practise them in his garden. Ibrahimovic revealed that he'd 'often spot solutions in games that I then played out in real life as a young player', while Bayern Munich and Germany defender Mats Hummels commented that 'maybe some people use what they learn in FIFA when they find themselves on a pitch'.

Not only does playing FIFA help modern footballers hone their technique and tactical awareness, it can reduce their stress levels, too. In 2015, an Iranian team of scientists led by Hamed Aliyari asked 32 20-year-olds to play FIFA 2015. Saliva samples were collected from the participants before and after competition to examine hormonal levels. This showed a drop in the stress hormone cortisol. So playing FIFA had eased the stress levels of the gamers. Reduced cortisol has also been shown to accelerate mental processing. Perhaps Pirlo had his PlayStation to thank for some of those magical, split-second passes that were his trademark.

This notion is not as far-fetched as it might sound. An earlier study (2013) conducted at Queen Mary University of London supported the idea that gaming can provide neurological benefits. The researchers found that individuals who played the game StarCraft displayed increased mental agility when it came to decision making and enhanced creative and lateral thinking. 'The work also revealed that real-time strategy games can promote our ability to think on the fly and learn from past mistakes,' explained study leader, Dr Brian Glass.

◀ The secret to former Chelsea player Cesc Fàbregas' success? Playing FIFA, of course!

who'd been out with a back problem. We had about 10 days and, as it transpired, seven hours to work with him. We tested him and, in particular, his turning ability, and noted that he was 0.67 seconds slower turning one way than the other. Their first game was against Portugal and he was marking Ronaldo. So that's a good 5m you're giving Ronaldo. We worked with him and brought the differential down to 0.32 seconds. That gave him confidence and made him quicker.' Antoniades doesn't name the individual but he would have been Icelandic as they drew 1–1 with Portugal, the eventual winners of the tournament, in their opening game.

Retraining footballing excellence

Altering an individual's biomechanics takes commitment and patience. It also requires the meticulous, innovative approach that's the hallmark of Ajax. In chapter 1, we looked at how the Dutch club are maximising GPS data with the Inmotio system. They've also recently partnered with Microsoft to pool all their analytics together, creating an accessible database that explicitly links different metrics, which makes spotting trends that much easier. For example, it might tell them that a left-back is being consistently beaten on the inside by a right-footed winger or a centre-back is standing in the wrong position to defend against strikers whose top speed is over, say, 35km/h.

But it's the biomechanical evolution unfolding in their miCoach Performance Centre that's taking the 'simple' actions of kicking, jumping and sprinting to a new technical level. With the aid of a series of cameras and sensors and a PhD student from a local university, the Ajax sports-science team are using biomechanical analysis to answer specific football-related questions. 'For example, in the past we've asked what's the best kicking technique for a goalkeeper as they can often suffer from groin problems,' says Ajax's head of sports science, Vosse de Boode. 'We'll analyse why some players suffer and others don't. Then we study the whole biomechanics of that kick to see if there's a noticeably different pattern between the two.'

The technology is still in its infancy and studies are sometimes affected by limited player availability, but the possibilities are endless. For example, if Ajax are scoring more from the left-hand side of the box than the right-hand side, is it because striker David Neres keeps slicing the ball past the left post when the ball is played from the right wing? If so, de Boode and her team could assess the biomechanics of Neres' landing foot. Does it leave him too cramped up? Does his right leg cut across the ball slightly because of lack of space? With the right staff and a commitment to change on the part of the player, issues like this could be solved within a couple of months, helping Ajax to maintain their position at the pinnacle of Dutch football.

Ajax aren't unique in embracing this in-depth biomechanical analysis but few are focusing on such specific performance-enhancing match situations. Take Manchester's Institute of Health and Performance (MIHP), which features a 16-camera motion-capture hall that measures aspects like ground-reaction forces and movements in a sportsperson's limbs. 'It's predominantly being used for rehabilitation rather than performance improvements but we haven't been open that long,' explains the centre's performance lead, Rachael Dawe. 'It might be an anterior cruciate ligament (ACL) issue and we can record information graphically and compare with a normal range of motion. Having the player's benchmark biomechanics to compare against is the ideal.' That's rare, Dawe concedes, as relatively few clubs are able to provide this – but a graphical reconstruction of the player still gives enough information to gauge where they are on the path to recovery.

Witness the fitness

An increasing number of English clubs are beginning to measure their players' biomechanics. Visiting Southampton's Staplewood training ground, it's noticeable that the gymnasium features a whiteboard that charts a series of tests to assess the players' capabilities as the season unfolds. There are sprint tests – 5m, 10m, 20m and 30m – all ticked off in the club's impressive training dome and logged by timing gates, as well as counter-movement jump-height and peak-power tests, both using a force

▼ Run specialist Mike Antoniades gave Iceland's defence a split-second advantage at the 2016 Euros, paying off with a 1–1 draw with Portugal

▲ Victor Wanyama undergoes a fitness test – specifically the counter-movement jump test – at the Spurs' training ground

platform to spring off and measure. The board also includes a player's maximum speed based on GPS readings.

These tests are a staple of the modern footballer's working life, not only helping clubs to track physical fluctuations during a season and beyond, but also giving the coaching team vital information to take on to the training pitch. 'That's where science and physiology comes in,' Southampton's director of performance science, Mo Gimpel, explains. 'Charlie [Austin] is not particularly quick at accelerating but his top-end speed is high, whereas Nathan [Redmond] is fantastic from a standing start but is middling for top-end speed. That sort of information can influence your tactics – if you're looking to play the long ball, say, Charlie's your man to run into the channels.'

Knowledge is certainly power when it comes to training, tactics and potential purchases, with the foundations of all three laid in the pre-season. Historically, this was the period where players, returning from their alcohol-fuelled holidays on the Costa Brava, would roll up to the training ground with a burnt face and bulging waistline. Southampton legend Matt Le Tissier confessed he often came back overweight and under-conditioned. It was hard to deny.

And it still happens, albeit it's no longer the norm. Then Newcastle manager Alan Pardew fined Hatem Ben Arfa £1,600 in July 2014 after the French midfielder arrived back from holiday 1.5kg over his target weight, while Pep Guardiola reportedly put Samir Nasri into a Manchester City 'fat camp' after he had piled on the pounds. But the damage had been done and Nasri was sold soon after.

The sports scientist is football's Big Brother, the all-seeing eye from whom there is no hiding place. This is something Cambridge United's Mark Roberts reflected on in a 2017 feature for *These Football Times* web magazine: 'As you approach the training ground, the butterflies usually kick in. The fitness tests will do that to you. Have you indulged too much and worked too little? Are you as fit as you were or as you need to be? Even when you're confident with your own physical state, you still don't know how you will compare with your teammates, because once the handshakes and hugs are over, the competition for shirts has begun. It's dog eat dog, survival of the fittest. Long gone are the days when players returned to training overweight or underprepared; sports science has seen to that...'

No one in the field has a keener eye than legendary fitness trainer Nick Grantham. His first job was head of sports science for British Gymnastics, based at Lilleshall National Sports Centre in Shropshire, which also housed the Football Association's School of Excellence between 1984 and 1999. Grantham then upped sticks to the English Institute of Sport before going solo. His eclectic freelance CV now includes stints with the Chinese football team and, more recently, West Bromwich Albion. Grantham knows what it takes to be an elite footballer and recognises the fact that football has had a fitness makeover these past few years.

'Players will now come back for pre-season training as fit if not fitter than when they finished the previous season,' says Grantham. 'The competitiveness of the Premier League and the commercial pressure of off-season tours mean you have to start day one with all cylinders firing.'

For lower-profile clubs like West Brom the pre-season schedule is not squeezed quite as hard by commercial demands as it is for global brands like Manchester United and Arsenal, who visited the US and China, respectively, in the summer of 2017. In fact, Tottenham Hotspur embarked on a *post*-season tour of South Korea and Hong Kong within hours of hammering Hull City 7–1 on the final day of the 2017 season. West Brom's pre-season fixture list was predominantly UK based with short trips to the Czech Republic and Spain to play Slavia Prague and Deportivo de La Coruña, respectively.

That's no criticism of West Brom and, like every Premier League club, once you take into account international duty many of their players might have as little as 10 days' rest before they're summoned back to training. 'A number of our players were involved in the European Championships in 2016,' Grantham explains. 'Pre-season for them was slightly later, which made it tricky to manage their workload alongside the rest of the playing squad.'

Long gone are the days when players returned to training over-weight or underprepared; sports science has seen to that...

MARK ROBERTS CAMBRIDGE UNITED

Whatever the length of summer break, Grantham at West Brom and many of his counterparts will have the players undertaking fitness tests at the end of the season and then on their return. Because they're fresher – and as long as they've spent some of their off season maintaining fitness rather than clubbing – the figures should roughly match. 'Pre-season is then about increasing fitness from that initial pre-season test,' says Grantham. 'It means you should grow stronger each year.'

How do you gauge how fit players are? There are myriad training tools available, with every club doing things slightly differently. Manchester City's world-leading training ground includes the Cybex isokinetic machine for neuromuscular testing as well as rehabilitation, while a sprint machine, which with its menacing-looking chains resembles something more attuned to torture, measures torque. The club

▶ Like all clubs, pre-season training at West Bromwich Albion is about setting benchmarks to build on

also uses the services of the Manchester Institute of Health and Performance, which sits within the Etihad campus and is part-funded by City as well as Sport England and Manchester City Council. Inside are computers, software and lab equipment to measure football-relevant physiological parameters like speed and endurance.

Keep returning to the Yo-Yo

No matter how wealthy a club is, arguably the most widespread fitness test is the Yo-Yo intermittent recovery test, developed by Danish exercise physiology legend Jens Bangsbo back in 1994. It's similar to the classic beep test that many of us suffered at school, but instead of running continuously back and forth between two cones – in the case of Yo-Yo, 20m apart – you 'enjoy' a brief pause between each shuttle. Hence the name. Like the beep test, the gap between beeps shortens, with the fittest running on for longer.

The Yo-Yo has usurped the beep test and become popular among football clubs because that stop-start nature is more representative of the way players exert themselves during a match. One study referenced by Bangsbo when he created the test showed that 16 marathon runners, who ran a pretty impressive average time of three hours and 12 minutes, scored below the level recorded by amateur football players. That's because of fitness specificity; in other words, while the elite runners

APPLY THE SCIENCE

Higher, faster, stronger

Follow these sessions by acclaimed fitness coach Nick Grantham to build your strength and conditioning in pre-season and maintain it throughout the season

Typically, pre-season strength-training sessions are longer and more frequent than the ones during the season. In fact, during the early weeks of pre-season it wouldn't be unusual for professional players to have an exposure to a strength stimulus every day. Once you're 'in' season, you'll have a maximum of two exposures each week and that could drop to one session during periods of heavy fixture congestion. With that in mind, here are two pre-season and two in-season sessions with benefits highlighted. Note that some of the exercises require basic gym equipment (barbells, dumbbells, etc.).

Pre-season session 1

Benefit: *strength development and force production*

1 **Squat:** 4 x 8 (i.e. four sets of eight reps)
Standing upright with knees slightly bent, and holding a barbell behind your head and rested on your shoulders, slowly bend your knees, ensuring your back remains as straight as possible. Return and repeat.

2 **RDL** (Romanian dead lift): 4 x 8
Holding a barbell with knees slightly bent and standing upright, keep back taut and hinge from the hips, keeping the bar close to the front of your legs as you lower it. Slowly return to start position and repeat.

3 **Hip thrust:** 3 x 10
Sit on the floor with your back resting against the edge of a gym bench and your arms stretched out to each side and resting on the bench. Your back should be straight and legs bent. Place a barbell – or get help from a friend or coach – across your lower abdomen. Simply thrust your hips up so that the barbell is now parallel with the height of the bench. Slowly return and repeat.

4 **Box jump:** 4 x 4
Having placed a foam plyo box around 15cm in front of you, adopt the squat position with feet around shoulder-width apart. Squat and explode up using your entire body including your arms. Land softly on the box on the balls of your feet. Step down and repeat.

5 **Band-resisted broad jump:** 4 x 5
Loop a gym elastic band around an immovable object at waist height. Put the other end around your waist. With chest out, knees slightly bent and feet hips-width apart, throw your hands down quickly and leap forward. Just be careful when you land as the band will immediately pull you back!

▶ S&C coach Nick Grantham works with WBA's Luke Daniels during pre-season

Pre-season session 2

Benefit: *functional loading (football-specific movements) and force application*

1 **RDL to box jump:** 2 x 5
Do the lift, then put the barbell down before going into the box jump.

2 **Single-leg RDL to box jump and horizontal plate push:** 2 x 5
The horizontal plate push involves pushing a barbell weight along the floor with your hands, taking small steps to avoid standing up.

3 **Box jump:** 2 x 5

4 **High knee switch with overhead press:** 2 x 5
Holding a dumbbell in each hand, alternate lifting above your head with right and left hands. At the same time, run on the spot focusing on getting your knees as high as possible. Your right knee should rise at the same time as your left hand, and vice versa.

In-season session 1

This session involves lifting relatively heavier loads at slower speeds to develop strength. For reference, in a Premier League week with a game on the Saturday but no midweek game, this would normally happen on a Tuesday.

Benefit: *strength*

1 **Squat or trap deadlift:** 3–4 x 5
A trap deadlift is identical to a normal deadlift but uses a trap bar. This is a barbell with a hexagonal enclosure in the middle, allowing you to stand inside it rather than holding the weight out in front as you would traditionally. This reduces stress on the back.

2 **RDL:** 3–4 x 5

3 **Vertical jump:** 3–4 x 3
Simply squat down and leap as high as possible into the air.

4 **Hip thrust:** 3 x 10

In-season session 2

This session involves lifting relatively lighter loads at higher speeds to develop the ability to produce force quickly.

Benefit: *power*

1 **Squat or trap deadlift:** 2–3 x 5

2 **Broad jump:** 2–3 x 3
Squat down and leap as far forwards as possible.

3 **Drop jump:** 3–4 x 3
Stand on a platform (e.g. a box), drop off, land briefly to absorb the shock and then immediately jump as high as possible.

had trained their systems to run at a consistently moderate pace, unlike the footballers they weren't used to running at a higher pace with rest periods in between.

As for player benchmarks, elite footballers will usually be able to run for 2,420m and above, which is considered a level 20.1, while 'moderate–elite' players will tend to reach 2,190m (level 19.3). Recreational players will register around 1,200–1,300m (level 16.3–16.5).

The reason marathon runners' and footballers' results differ so markedly is to do with energy systems. While marathon runners rely heavily on aerobic conditioning, developing a cardiovascular system that efficiently burns fat and delivers oxygen, footballers draw on a mix of aerobic and anaerobic energy. 'You have three energy systems – two anaerobic, one aerobic – and duration dictates which one dominates,' says Grantham. 'The first is your ATP-PC system [fuelled by adenosine triphosphate (ATP) and phosphocreatine (PC)]. It produces enough energy for around 10 seconds of really hard effort. Think the 100m. After that you go into another anaerobic energy system. It can produce enough energy without oxygen for about a minute or 90 seconds before burning out. Think the 400m. Any effort over that duration becomes predominantly aerobic in nature, which means it uses oxygen.'

▲ Jens Bangsbo (right), leading Denmark's Nicklas Bendtner, is the father of the Yo-Yo Test

Because football is a mix of high-intensity running, all-out sprints, jogging and walking, it works all three energy systems. This provides fertile ground for exercise physiologists, especially Bangsbo, who has written books specifically on football and energy systems.

'Despite players performing low-intensity work for more than 70 per cent of the game, heart-rate and body-temperature measurements suggest that the average oxygen uptake for elite football players is around 70 per cent of the maximum, which is pretty high,' explains Bangsbo. 'That's partly explained by the 150–250 intense actions a top-class player performs during the game.'

Bangsbo investigated footballers' anaerobic energy demands by taking muscle biopsies of Danish third-division league players before and after games. 'You could measure how much lactate was in the muscle, which is a by-product of producing energy without oxygen [anaerobically]. After the game it was pretty high…' This provided biochemical evidence of footballers' anaerobic energy needs.

Aerobic metrics like heart rate and body temperature can be measured throughout the game, but this is not possible for lactate – it's neither practical nor desirable (from the players' point of view) to have muscle extracted while they are bearing down on goal! This means that it is hard to know the exact breakdown of energy use from the aerobic and two anaerobic systems at any given point in a match. However, it was clear to Bangsbo that all three systems were significant and needed to be taken into account in training.

Building speed-endurance

As part of his research into energy systems, Bangsbo developed a programme of speed-endurance training. 'We did a study with FC Rosengard, formerly known as LdB FC Malmö, who play in the Swedish women's premier division. We reduced training volume by about 17 per cent but introduced more high-intensity aerobic training and speed-endurance training. And what did we see? By the end of the season, their Yo-Yo performance had risen from 1,600m to 2,100m. Is that an advantage on the pitch? Yes, it is. Injuries also dropped by 60 per cent, which meant more players were available. That season we won the championship and qualified for the Champions League.'

▼ Bangsbo developed a ground-breaking speed-endurance programme for FC Rosengard and their then star player Marta Vieira da Silva

This focus on speed-endurance training involves sessions at high intensity and with long recovery periods to stimulate all three energy systems. The players follow it two to three times a week, 'but you must remember to include the ball for motivation. This speed-endurance work is something we did at Juventus, but we'd have to be creative. You might have specific drills with players like [Alessandro] Del Piero and [Zinedine] Zidane, where one would get the ball, turn around, give it back and play it through to the player to shoot. We'd measure it so that it might be 23 seconds of hard work followed by two-and-a-half minutes of rest. It met not only the technical demands of matchplay, but the physiological ones, too.'

That mix of aerobic and anaerobic energy supply is often weighted more towards aerobic sessions when players hit the off-season. 'This is because it lays a physiological foundation to build anaerobic work on,' says MIHP's performance lead, Rachael Dawe. 'It won't necessarily improve your sprint prowess but it'll help with repeated sprints.'

Managing the amount of aerobic and anaerobic workout is a balancing act; each club follows different criteria. According to Southampton's head of sports science, Alex Gross, players will return from their holidays with a VO_2 max (measure of aerobic capacity – how much oxygen a player can process and use to fuel muscles each minute; see box overleaf) at an average of 62ml/min/kg. 'It's not up there with elite-level cyclists [who register 80-plus ml/min/kg because of the predominantly aerobic nature of their sport], but it's not bad,' he says. 'Sometimes we'll do a test here where we'll take blood and measure things like lactate every three minutes. If we want to do a full-on gas-analysis test, we'll go to Solent University.'

◀ Southampton U23 player Jake Flannigan undergoes a sprint test

The lactic test

THIS OFTEN-USED TEST MEASURES SPEED-ENDURANCE CAPABILITY

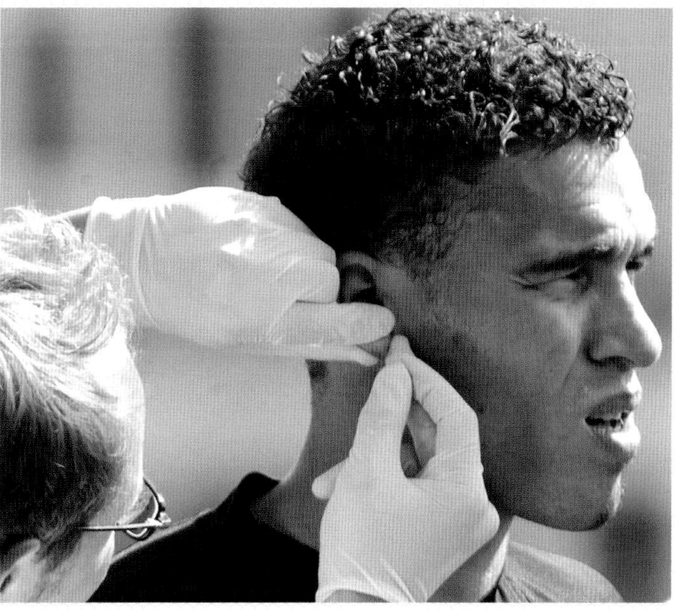

In July 2017 Liverpool offered a peculiar insight into the life of a professional footballer by posting a video of a pre-season lactate run on their official Twitter feed. 'Start off nice and slow at 8km/h,' guided the sports scientist, 'and increase by 2km/h each lap.' After every lap, the club's team of sports scientists and physiologists also took a sample of blood from the players' earlobes with a Biosen C-Line analyser to measure lactate levels. If the level of lactic acid in their blood exceeded a certain amount, they were out of the drill.

Testing and training to improve lactate threshold is one of the most common and effectively used performance markers in football, the aim being to identify the highest intensity at which a footballer can train or compete before hitting the wall because of high levels of blood lactate.

Jürgen Klopp and his team know that one of the keys to success is to increase sustainable power and speed while playing at the lactate threshold. This is defined as the 'exercise intensity at which the blood concentration of lactic acid begins to exponentially increase'. Lactic acid is a by-product of breaking down glucose for energy. It's recycled to produce more energy but, as intensity increases, it can't be recycled quickly enough. Lactic acid builds up in the muscles, leading to more acidic blood and eventually fatigue.

Arguably, for footballers the lactate threshold is a more important metric than the VO_2 max. Whereas VO_2 max is an indicator of basic endurance, the lactic test shows the intensity a player can work at. The higher your lactic threshold, the more repeated sprints you will be able to do.

Interval training and a healthy lifestyle helps raise lactic threshold, so it comes as no surprise that workhorse midfielder James Milner won the test. Known as one of the most dedicated players around, Milner's never touched a drop of alcohol in his life.

▲ The lactic test is a proven and relatively simple method to measure anaerobic capacity

Breathe deep, run fast

VO_2 MAX IS A MEASURE OF AEROBIC CAPACITY, ONE OF THE KEY ASPECTS OF PHYSICAL FITNESS THAT ENABLES PLAYERS TO MAXIMISE THEIR TALENT

Southampton's head of sports science, Alex Gross, stresses the effect of VO_2 max on performance. But what exactly is it? Essentially, VO_2 max is the maximum amount of oxygen a person can utilise during exercise – the higher the better as muscles burn oxygen for fuel. It is expressed as millilitres of oxygen used in one minute per kilogram of body weight: ml/kg/min.

'What we've found with VO_2 max is that it's very much player dependent,' says Tony Strudwick, Manchester United's former head of performance, now head of performance with the Welsh national team. 'Wide players and central midfielders record the highest numbers. But we've had some top players whose VO_2 max measured below 60ml/kg/min (recreational athletes tend to score around 45–50ml/kg/min). Some players have hit the 70s. That's nothing compared to professional cyclists, though, who are in the 80s and even 90s.'

The reason that footballers' VO_2 max scores aren't as high as those for endurance athletes like cyclists and marathon runners is that football has a strong anaerobic component, too. 'But it still impacts performance because there's a relationship between VO_2 max and repeated sprint ability,' Strudwick continues.

Several studies, notably one conducted in 2004 by Professor David Bishop, now of Victoria University, Melbourne, have shown that a higher VO_2 max delays fatigue during repeated sprints as it increases the ability to tolerate, remove and buffer hydrogen ions from anaerobic work (which produces huge swathes of lactic acid).

Using match-analysis technology, Jens Bangsbo found that a sample of international players performed 28 per cent more high-intensity runs and 58 per cent more sprints than professional players of a lower standard. Cristiano Ronaldo is believed to possess a VO_2 max of 75ml/kg/min. There's no reason to doubt that reading when you see him sprint like a cheetah from the halfway line. Still, whatever the Portuguese galáctico's fitness data, tactics can act as a safety harness. 'If you're Ronaldo, you pick your moments to sprint and make sure you have someone like Wes Brown or Darren Fletcher behind to support you,' says Strudwick.

The most accurate method of measuring VO_2 max is by means of a gas-analysis test. Wearing a mask over the mouth and a heart-rate monitor, the athlete will start running on a treadmill at low intensity with the speed increasing every two minutes. The mask collects expired air to measure gas concentrations and determine the maximal volume of oxygen, measured in litres of oxygen per minute, that the body is utilising.

▲ According to Tony Strudwick, wingers and central midfielders recorded the highest VO_2 max scores

> A warm-up should be developed for *what you're about to do*, so you shouldn't be doing the same thing every time.
>
> **MO GIMPEL** DIRECTOR OF PERFORMANCE SCIENCE, SOUTHAMPTON

The club's six-week pre-season programme starts with two weeks of slow, steady-state running before increasing the intensity. 'We've shown that if we can raise someone's VO_2 max, they have less chance of being injured throughout the season,' says Gross. 'Basically, we want to shift the lactate curve right so they can work harder and longer.'

From experience and analysing the literature, Gross then adopts a mixed-model approach that's similar to Jens Bangsbo's model. 'The players might run 85m in 15 seconds, have 15 seconds rest and then repeat that eight times. And they might do three blocks of that. There's research that shows doing that sort of thing twice a week raises your VO_2 max.'

It's taxing but, as Nick Grantham mentioned earlier, players are returning to pre-season in much better shape than in years gone by – not wanting to fall foul of the all-seeing sports scientist. 'The players will train while they're on holiday,' says Gross. 'Generally, I get in touch with the hotel where they're going on holiday, find out what equipment they have and, even if there's only a rowing machine or exercise bike, you can still give them a session.'

The modern footballer has to accept this 365, 24/7 monitoring. Would you put up with your boss ringing you on holiday every morning to check you were reading your work emails? Perhaps you would if you were being paid £100,000 or more a week.

Getting warmer

Physical preparation is clearly an all-year-round affair with not much downtime. And a crucial part of the process, each and every session, is the warm-up. Head to the nearest park on a Sunday morning for local football, and you'll see thousands of boys, girls, men and women pulling the old hamstring stretch and undertaking a few jogs and sprints before asking their mates about the previous night's *Match of the Day*.

The warm-up in professional football, both at senior level and in academies, is very different. As I watch Southampton's youngsters prepare to face Spurs in an under-18 match, the warm-up lasts 35 minutes, culminating in a small-sided game between Saints players who appear to be working a lot harder than their opposition.

'I could be completely wrong but ours looks like a science-led warm-up and Spurs' is a coach-driven warm-up,' Southampton's director of performance science, Mo Gimpel, observes. It is true that the Spurs players seem to be focusing on set plays, in contrast to the high-tempo work of Southampton. 'The game is going to start at a high pace, so our guys need to be ready for that. So one touch, multiple changes of direction ... that's the game. That's a training session in itself.

'A warm-up should be developed for *what you're about to do*, so you shouldn't be doing the same thing every time,' Gimpel continues. 'If your training day's going to involve a lot of pressing, deceleration and acceleration, small-sided games are good. But if you're going to be really sprinting over long distances, your warm-up needs to cater for that with longer warm-up sprints. And your pre-match warm-up should be dictated by the tactics you're going to use in the match.'

As it transpired, Spurs won 2–1, highlighting that a tactical, skilful and dynamic sport like football doesn't always conform to textbook sports science. Then again, although Southampton lost on the day, perhaps their warm-up offered better protection against injury in the long run than that of Spurs. And perhaps Spurs might have won by more than one goal had Southampton warmed up differently.

Whatever the content of the actual warm-up, all professional players start by doing pre-activation work. This consists of a set of stretches, balancing movements and bodyweight exercises focusing on football-specific muscles (e.g. hamstrings and calves) to prime the neurological system for upcoming activity.

At Southampton's training ground, they often do this in the club's gymnasium upon foam flooring that's designed to improve proprioception (awareness of body in space). Then they head outdoors for the main warm-up and training. 'That was brought in by Mauricio Pochettino [Southampton manager, 2013–2014] to improve sprint work while reducing impact,' says Gimpel. 'Though we use it for pre-activation work, we don't use the gym as much for sprinting under the new regime so we're considering going back to a traditional rubber floor.'

When it comes to match day at St Mary's, the Southampton players will do pre-activation work for 15 minutes in the changing room before heading out to the pitch. 'It's a club-wide policy,' says Gimpel. 'If you're going out to train or play, you must do a pre-activation session. Take exercises like the pelvic thrust. It's pretty well established that if you can control your femur in relation to your pelvis and lumbar spine, incidence of injury drops massively.'

Once the players are 'pre-activated', they'll head out to the pitch and perform a series of routines designed to mimic match conditions. So a centre-back might play long balls to a fellow centre-back as that's a common pass for many pairings, while midfielders might play short, sharp passes to each other.

Professional warm-ups are complex affairs but they're essential in preparing the cardiovascular and respiratory systems, and muscles, for the match. It's long been established that raising muscle temperature improves performance; a classic 1987 paper in the *European Journal of Applied Physiology* by Professor A.J. Sargeant showed a 4 per cent increase in power output per 1°C rise in muscle temperature. A 3–4°C increase is deemed optimal. Hence the 35-minute warm-ups.

FIFA place such importance on warm-ups in football, especially at youth level, that they've created the FIFA 11+. This 15-stage routine – consisting of running, strength, plyometrics (explosive activities to increase power), and balance exercises – is claimed to cut injuries in youngsters by up to 45 per cent and reduce the severity of injuries by up to 30 per cent.

'It's a useful tool,' says Arsenal Women FC's lead physiotherapist, Sam Blanchard, who formerly worked with Brighton's men's team, 'though in all honesty, most clubs follow a more sophisticated routine. Where the 11+ has come into its own is in less mature footballing regions like the Middle East and Africa and also in women's football, which historically don't have the sports-science support of the men's game in the major leagues.'

Blanchard also stresses that it's extremely useful for amateur footballers. So if you play or coach football and you're doing the old-fashioned lunges for three minutes before kick-off, take a look at the FIFA 11+ online. The science proves it works! But whether you're John Stones or John Smith, and no matter how many sports scientists are involved in football in the 21st century, the warm-up is still not perfect. One recent study focusing on the Premier League and Championship found that there was on average a 13-minute gap between end of warm-up and kick-off, potentially undoing the good work of the (on average *31-minute-long*) warm-up.

Of course, Mike Antoniades' quest to extract extra speed from Hoffenheim players and Vosse de Boode's technique-refinement work at Ajax are both hampered if the elite footballer is worn out from fixture congestion. With international players ticking off up to 70 games a year, it's a very real threat. Which is exactly where our next chapter comes in – just how does the modern footballer recover in time for the next match?

▼ The Man Utd squad undergoes pre-activation work on their 2018 pre-season tour of the US

KING
POWER

RAPID RECOVERY

Cristiano Ronaldo is a man whose impeccable torso and picture-perfect abs look like they've been chiselled from a block of ice, which is not so far from the truth. Back in 2013, the Real Madrid legend invested some of his gargantuan wages in a cryotherapy chamber – also known as an 'ice box'. Fitted in his £4.8 million Spanish home, this unit is designed to improve recovery and ensure he lines up at the next match in optimum condition.

He's not the only one utilising this cutting-edge technology. Leicester City attracted attention for the remarkably low incidence of injuries sustained during the club's Premier League-winning 2015–2016 season. Their sports-science department credited the cryotherapy chamber they had installed as one of the secrets of the Foxes' memorable success.

Sam Erith is Manchester City's head of sports science. The Cornishman is one of the game's most respected sports scientists and formerly led Tottenham Hotspur's programme. On a sunny Friday morning in August, just prior to the players flying down to Bournemouth for a Saturday lunchtime kick-off – City would win 2–1 after an explosive end to the game that saw Raheem Sterling sent off after his late winner – Erith showed me the facilities that are the envy of leagues around the world.

◄ Leicester's sport-science department – and the cryo chamber – received much credit for their 2015/2016 league-winning season

'This is the cryotherapy unit,' he says. 'We tend to use cryotherapy when we have a number of games in close succession; the players will come in the day after a game.' Plastered on the crisp white walls is a giant poster educating the players about cryotherapy with the headline grabber: 'The quicker you recover, the harder you can train...' Dramatic pause. '...The harder you can train, the better you'll be.' Beside the slogan are the physiological benefits: '1 Enhances sleep quality and mood; 2 Helps control inflammation; 3 Reduced muscle soreness and fatigue; 4 Restores running performance quicker.'

We'll delve a little deeper into each shortly but inescapable was the sartorial checklist: face mask, elbow sleeves, shorts, gloves, knee sleeves, socks and, the ultimate attack on a fashion-savvy footballer's dignity, Crocs. 'That's all to protect the players' extremities,' says Erith. 'It's needed as temperatures reach minus 140°C.'

Chilling warm-up

Manchester City's set-up actually comprises two walk-in chambers, so that once the players are dressed up in their rather frightening attire, they enter chamber one, which hits the relatively tropical heights of minus 50°C. 'That's the preparation phase,' Erith explains. 'After 30 seconds, an alarm goes off and lights flash, signalling for the player to exit that chamber and enter the next one. That's set to minus 140°C. They'll then stand in there for two minutes before exiting the chamber and warming back up to room temperature.'

It's extreme but, as per the poster, there are numerous mooted benefits, starting with enhanced sleep quality. It's been shown that your body releases the hormone nor-adrenaline and activates the neurotransmitter acetylcholine when exposed to cold temperatures, both of which are important to the sleep cycle. Cold also stimulates endorphins, the happy hormones that help to improve your mood. A further study showed that it also reduced levels of inflammation markers – desirable for swift recovery – via increasing cytokine levels in the blood.

And that's not all. A 2010 paper in the journal *Sports Medicine*, entitled 'Whole-body cryotherapy in athletes', reported that the therapeutic effects related to changes in muscular enzyme activity – specifically reduced creatine kinase and lactate levels, again both commonly related to high levels of inflammation.

The idea of super-freezing the body to aid recovery goes back to the late 1970s when Japanese scientists touted it as a potential method to relieve joint pain in patients with multiple sclerosis or rheumatoid arthritis. Now, high-tech chambers chilled with liquid nitrogen are everywhere. And that's despite a significant number of studies showing they have no effect at all, with critics of the research in support of cryotherapy also arguing that methodologies are too inconsistent and subject numbers too low to draw any robust conclusions.

Take a 2017 study by Professor Liam Kilduff of Swansea University, who studied 14 Premier League academy players after a single cryotherapy session (minus 135°C for two minutes), undergone within 20 minutes of a repeated sprint exercise (15 × 30m). Although the players showed increased salivary testosterone – an indicator of recovery – the treatment had no effect on levels of cortisol, blood lactate or creatine kinase, nor on performance (peak power output) or soreness perceptions.

Admittedly, 14 is a small sample size but the Cochrane Review – the gold standard in healthcare evidence – pooled the results of four previous studies involving 64 physically active adults and concluded that there is 'insufficient evidence to support use [of cryotherapy] to relieve muscle soreness after exercise'.

'It's true that the research is equivocal,' says Erith, 'but, anecdotally, it seems to work.' It's why the Manchester Institute of Health and Performance, a state-of-the-art facility in the grounds of Sportcity, are undertaking a three-year study into cryotherapy with the aim of producing a new gold standard. The results are due at some point in 2019.

That research could see new protocols being adopted, with cryotherapy sessions two or even three times a day, or for longer after every game. But many clubs require further evidence before spending a cool £100,000 on their own chamber.

'Yes, an increasing number of clubs are using cryotherapy for recovery and many of our players have used it on international duty [the FA base at St George's Park in Staffordshire doesn't have a chamber but they've been known to ship in a mobile unit] and said it's amazing, but it's a heavy initial cost plus a couple of grand a month running costs for nitrogen, so is this something we should invest in or not?' muses Southampton's director of performance science, Mo Gimpel. 'Some clubs will just buy one because they don't have to worry about money, but that can make you complacent – even lazy – because you're not really invested in it and so you end up not using it.'

They're business people and won't just give us the money – we have to prove it works.

MO GIMPEL DIRECTOR OF PERFORMANCE, SOUTHAMPTON

Instead, Southampton plan to rent one for a month and engage with a physiology expert to set an 'academically sound research protocol' and then evaluate its effects. 'We'll talk with the lads before, during and after, and assess the impact, too. Of course, the results might be positive but be purely down to the placebo effect. But that's fine – placebo's huge and something we use all the time.'

At the end of the month, if convinced by the merits of cryotherapy, Gimpel will make a presentation to the board requesting the money. 'They're business people and won't just give us the money – we have to prove it works.'

As it transpired, Southampton's ambitions grew beyond a small-scale study. In October 2018 the club put an advertisement on the British Association of Sport and Exercise Sciences website seeking a PhD student to study 'The impact of cryotherapy exposures applied across the training cycle of professional soccer players' in collaboration with St Mary's University, London. With the course running three years, firm conclusions will not be revealed until the 2022–2023 season.

Cool immersion

Back in August 2017, club owner, Katharina Liebherr, sold 80 per cent of her stake in Southampton to Chinese businessman Jisheng Gao for £210 million. That's serious money but, as highlighted, Gao might need some convincing to spend even more on cryotherapy units. Instead, Gao should digest a 2014 study by Dr Chris Bleakley of Ulster University. Bleakley analysed ice, cold-water and whole-body cryotherapy studies carried out by himself and other researchers and found that in fact the lowly (and incomparably cheaper) ice pack delivered the greatest reductions in skin and intramuscular temperature, theoretically stimulating accelerated recovery.

Why might this be? Although the temperature in a cryogenic pod is much colder than ice, the cold from ice applied directly to the body has a greater chance of penetrating through layers of skin and fat to reach the target soft tissue.

▲ Ice baths are commonly used for both cooling and recovery. Aaron Ramsey used it for both on a pre-season tour of China

It's why clubs continue to use ice baths as a recovery aid. These are somewhat misleadingly named, as they contain cold water not ice. 'If a player is reporting soreness and we know there's some inflammation in there, an ice bath can reduce some of the markers,' says Robin Thorpe, head of recovery and regeneration at Manchester United. 'We have two ice baths at the club, both purpose-built with jets. We can change the temperature from 2°C to 20°C and usually the players will be in there for around 10 minutes. There's plenty of research that shows it significantly reduces bloodflow to the muscles, which is what we're after [because it reduces inflammation and swelling].'

Thorpe suspects that players at other clubs might warm themselves back up under a hot shower immediately after getting out of the ice bath, but he won't allow United players to do that, because it reduces the recovery effect of limiting blood flow to the muscles.

Players like Paul Pogba and Marcus Rashford are happy to stoke the Schadenfreude of their combined 8 million Twitter followers by posting snapshots of themselves shivering their way through the congested Christmas programme. It gives the impression that ice is the number one recovery tool. But this is not so, according to Thorpe. It's a tricky balance. 'Ice baths can actually blunt physiological adaptation [see box]; there's certainly some evidence of that in strength-training studies. That's

> If we're only playing once a week, we'll probably avoid the ice bath because the players will have time to recover naturally.

ROBIN THORPE HEAD OF RECOVERY AND REGENERATION AT MANCHESTER UNITED.

why we do it on an individual level. It's also why if we're only playing once a week – admittedly that's rare – we'll probably avoid the ice bath because the players will have time to recover naturally.'

It's why CryoSpas, who kitted out Manchester United at Old Trafford and their Carrington training ground, recommend: 'In the competitive phase of the season the use of ice baths will help minimise fatigue and aid recovery, thereby improving performance and lowering the risk of injury. In the pre-season, or power-building, phase of training the use of ice baths may adversely affect the adaptive response.'

A 2015 study led by Professor Lilion Roberts of the University of Queensland, and featured in the *Journal of Physiology*, had 21 men undertake a 12-week strength-training programme. One group followed up with 10 minutes of cold-water immersion; the other with 10 minutes' active recovery (a low-intensity and low-impact exercise like indoor cycling). The result? The active-recovery group increased their strength and muscle mass more than the cold-water group. The researchers concluded that this was down to the cold water hampering the activity of the enzyme creatine kinase, which regulates muscle growth.

'On the other side, there's research that suggests cold-water immersion "improves" an aerobic gene,' adds Thorpe. That research includes a 2017 study by Nana Chung, published in the *Journal of Exercise Nutrition and Biochemistry*, which showed that cold-water immersion stimulates the expression of genes related to mitochondrial biogenesis. This term is a bit of a mouthful, so we will save it for the chapter on nutrition (chapter 5).

Just to sprinkle further uncertainty into the mix, Southampton's director of performance science, Mo Gimpel, suggests that there can be an adverse psychobiological response to cold-water immersion, particularly among some players raised in hot climates. These players found the ice baths so traumatic that it led to an increase in the stress hormone cortisol. 'That's exactly the opposite of what we're after because it impedes recovery,' says Gimpel.

Confused? Clearly the research findings for cold-based recovery, whether in ice baths or cryotherapy chambers, are mixed. It is to be hoped that future research, focusing on specific

Physiological adaptation

In broad biological terms, a physiological adaptation is any internal change within the body in response to an external stimulus in order to help the body maintain homoeostasis (equilibrium). Physiological adaptations that footballers will experience in response to training include muscle growth and raised VO_2 max and lactic threshold (see pages 51–52).

factors like body composition, genetic disposition and physical size – all of which could affect how a player physiologically responds to low temperatures – will be more conclusive.

Markers of fatigue

'There are various hormonal markers we'd examine throughout the season,' says Adam Brett, head of medical services at Brighton and Hove Albion, when I ask him how the Premier League club measure for fatigue. 'One of the key hormones we look at is testosterone – low levels are a sign of heavy fatigue. We also used to focus on creatine kinase.'

Creatine kinase (previously mentioned in relation to cold-water immersion) is an enzyme in your blood that your muscles need to function. A high level is often taken as a sign of heavy fatigue. But, says Brett, this might not be the case. 'What you find is that when you play traumatic levels of sport, creatine kinase undergoes a steady escalation; it never resets to level one. I also discovered from working in rugby that micro-contusions [tiny tears in the muscle] would elevate levels. That's why using creatine kinase as a marker of fatigue can be misleading.'

Brett's conclusions are supported by 2012 research in the *Journal of Nutrition and Metabolism* by Marianne Baird of the University of the West of Scotland. Referring to creatine kinase, Baird stated that there 'is indeed controversy in the literature concerning its validity in reflecting muscle damage as a consequence of intense exercise'. She hypothesised that differing creatine kinase levels could be down to a plethora of factors, including disruption in energy flow through the body, neurological changes and even ethnicity.

Brett concedes further research is needed, and he could well undertake it at his club's impressive £32 million training ground, the Amex Performance Centre, which includes cutting-edge medical and gymnasium facilities. But it's not just the Premier League clubs that are upgrading their training grounds in search of success. League One Fleetwood Town have invested £8 million in their new centre, Poolfoot Farm, and fellow League One side Bristol Rovers are planning to build a performance centre just off the M5 motorway.

In contrast, Plymouth Argyle players train within a stone's throw – with a good arm – of their stadium, Home Park. This is Harper's Park, which according to the club's website has the following facilities: grass pitches. And that's it.

What they lack in resources they make up for in curiosity – and it was this spirit, and a unique service, that helped them win promotion to League One in 2016–2017. 'We tested the players throughout the season and were pleased with the results,' says Dr Jan Knight, founder of local company Knight Scientific, who work closely with Plymouth Argyle, and have developed a test that can help to prevent overtraining syndrome. 'Using the test there were 9 per cent fewer training days missed the next season, 12 per cent fewer due to muscle issues and 22 per cent fewer due to illness.'

Oxygen boost

FOOTBALL'S BIGGEST NAMES ARE SEEKING PEAK PERFORMANCE VIA PRESSURISED OXYGEN

An eagle-eyed photographer from Spanish sports daily *Marca* drew a huge response from the Twitterati in July 2016 after snapping five-time World Footballer of the Year Cristiano Ronaldo exiting a dental clinic while on holiday in Ibiza. Reportedly, the Portuguese maestro wasn't there for work on his molars; instead, he was taking advantage of their side project – a hermetically sealed hyperbaric chamber. But why?

'Hyperbaric Oxygen Therapy (HBOT) is a non-invasive way of administering pure oxygen at twice the normal atmospheric pressure inside a private, purpose-built chamber,' explains Rob Pender of London-based Hyperbaric Oxygen Therapy Centre. 'This enables the body to take in 10–15 times more oxygen than normal.'

This extra oxygen hit, according to Pender, accelerates recovery from injury by relieving inflammation and thus making it easier for oxygen to get to the injured area. 'It's been recorded that ligament and hamstring tears heal at a faster rate – up to 38 per cent quicker,' he says.

HBOT is also used to recover from fatigue, the idea being that increased oxygen supercharges many of the body's metabolic processes, from clearing out free radicals (unstable atoms believed to cause cell damage) to producing more energy. According to a 2011 study by Mariana Cervaens of Fernando Pessoa University in Porto, Portugal, the scientific evidence in support of HBOT as a treatment for professional sportspeople is promising, but more rigorous and larger-scale trials are required before any firm conclusions can be drawn. Critics of HBOT argue that the body has a plentiful supply of oxygen to recover efficiently from injury and fatigue in most situations – apart from at high altitudes.

'We can't say who, but we worked with a Premier League club who'd struggled all season and were sitting at the foot of the table,' Pender says. 'They weren't a poor side but seemed to lose silly goals or missed chances in the final 15 minutes of each game. The conclusion was the players were fatigued.

> This extra oxygen hit accelerates recovery from injury by relieving inflammation and thus making it easier for oxygen to get to the injured area.
>
> **ROB PENDER** OF LONDON-BASED HYPERBARIC OXYGEN THERAPY CENTRE

'It was decided that three key players – a central defender, central midfielder and striker – would undertake HBOT for two days a week, with two 60–90-minute sessions per day. This continued for three weeks. They were "discharged" with eight games remaining because of shortage of time. The results were amazing – they won six, lost one and drew one of those last eight games and avoided relegation. The following year the sessions were repeated regularly ... and they were crowned champions.'

How much of an impact the oxygen boost had on the unnamed team – Leicester! – is impossible to quantify, but that has not stopped elite footballers and clubs from queuing up to go hyperbaric.

▲ Graham Carey was one Plymouth Argyle player who benefitted from the ABELsport test, helping the club to promotion

The company's ABELsport finger-prick blood test, Knight explains, can detect infections, sometimes up to 48 hours before any symptoms appear. It's that sort of statistic that has seen Dr Knight's test attract the attention of Premier League clubs like Arsenal and Newcastle. Identify fatigue or illness early and the coaching staff can ease the player's training load. So how does it work?

'First, you must dilute the blood in a test tube,' explains Knight. 'You then "activate" the white blood cells in the diluted blood by injecting fMLP. This is a small protein molecule derived from bacteria that binds to the white blood cells and activates them to produce oxidants. These oxidants react with Pholasin to produce light.'

Pholasin, a registered trademark of the company, is the most sensitive detector in the world of a particular oxidant called superoxide. When an infection enters the body, or muscles become fatigued through overtraining, the body responds by raising levels of superoxide. This in turn leads to the production of other oxidants collectively known as reactive oxygen species (ROS). Pholasin reacts with these ROS to emit light, which is detected in a portable instrument developed by Knight

Scientific called the ABELmeter. The readings can be plotted on a curve, from which the state of a player can be determined. It is the extreme sensitivity of Pholasin to this metabolic early warning sign that makes the test so effective.

Every week, Knight took finger-prick blood from the squad of 25. 'From that we could pick up various degrees of muscle inflammation, measured by extremely high light signals after exercise,' says Knight. 'It also gauges how well they've recovered from a match and as a way to assess fatigue ... we had a good season all-round.'

Heart-rate variability training

HRV, or heart-rate variability training, is used by Manchester United and AS Roma, among a number of top clubs. Heart rate monitors are a staple in the athlete's training toolkit – but what exactly is HRV training?

'The idea is that small variations in the beat-to-beat timing of the heart reflect the body's level of stress,' explains Simon Wegerif, founder of ithlete, a company that specialises in HRV training tools. 'Greater variations between beats – an increase in HRV – is associated with parasympathetic activity (rest and recovery); a reduction in the variations between beats – a decrease in HRV – is associated with sympathetic activity (fight or flight).' The key here is to understand the distinction between heart rate, which is the number of beats per minute, and HRV, which is the degree to which the interval between individual heartbeats varies. 'Hence, if you wake up one morning and HRV is very low,' Wegerif continues, 'that could mean you're neurologically fatigued. So if you had planned a high-intensity training session, you might decide to do a recovery session instead. And vice versa.'

The idea is to prevent overtraining, optimise recovery and therefore elicit peak performance. The ithlete technology works by attaching a finger sensor or chest strap monitor that sends information to your smartphone. Recordings must be

The autonomic nervous system

The parasympathetic and sympathetic nervous systems are sub-divisions of the autonomic nervous system, which acts largely unconsciously to regulate a variety of bodily functions such as heartbeat, blood flow, breathing and digestion.

The sympathetic division prepares the body for intense physical activity and is often referred to as the fight-or-flight response. For example, it causes the heart rate to accelerate and eye pupils to dilate. The parasympathetic system, on the other hand, is a far more chilled-out division, doing almost entirely the opposite; in other words, it relaxes the body and slows down the heart rate. For this reason, it is sometimes referred to as the 'rest and digest' system. There's also a third part of the autonomic system known as the enteric nervous system. This deals with activities of the gastro-intestinal tract.

taken immediately after waking to establish consistent readings. After a couple of weeks you'll start to see patterns, which you can interpret, via a series of graphs, to determine your physiological state to train.

HRV training can combine with other metrics to give the practitioner a deeper understanding of a player's fatigue state. 'As well as the chest strap [to measure HRV], we use EEG to measure brain activity,' says Dave Tenney, innovative sports scientist formerly of Major League Soccer side Seattle Sounders, now of Orlando Magic basketball team. 'Pairing the EEG with the traditional HRV data is powerful.'

An EEG, or electroencephalogram, is a test to track electrical activity in the brain by means of small sensors attached to the scalp for about 30 minutes. It all sounds a touch Frankenstein, but there is a performance rationale. In high-intensity sessions, the voltage of your brain decreases; essentially you lose charge. If your brain voltage drops too low this can lead to a state known as central nervous fatigue, which is a temporary inability of the brain to produce enough electricity to function optimally. Omegawave, a leading Finnish provider of HRV-training equipment, has shown, for example, that after a long and exhausting training session the brain voltage of a world-class tennis player dropped from 13 to minus 28 millivolts. If EEG readings remain low, it's a sign of overtraining and that the player needs rest.

'Here's how the EEG element of HRV works in a session with us,' Tenney begins. 'I'll have a coaches' meeting at 10 a.m. to discuss what loading we're looking for that day and what drills and sessions will elicit that loading. At the same time, our

◀ AS Roma have used HRV to influence workload for years

interns may be doing Omegawave assessments and retrieving fatigue questionnaires from players who've recently played, and will shoot me those results while we are in the coaches' meeting. Getting this info "in real time" can be a huge benefit in the planning stages, as you get a sense of the overall fatigue level of the group, which is important in such a long season.

'Once training starts, I will often take the first 25–30 minutes of the session, involving a dynamic warm-up. We'll then move on to a passing exercise aimed at preparing the athletes physically for whatever that day's target may be. If the EEG scores came in OK, I might continue with high-velocity work or some bouts of speed-endurance. If not, we might ease off with some easy running with the ball.'

EEG is insightful, innovative and massively impractical. That's why EEG readings are taken only occasionally, says Tenney, as are standard Omegawave HRV readings. 'The ideal is that players take a daily reading first thing in the morning but it's unrealistic that they'll do that so we tend to take measurements two to four times a week.'

If HRV training is to be accurate, readings must be taken consistently, ideally when the player is in a rested state. In the dynamic world of professional football, that's not always possible – and so support staff have to be creative. 'I've discovered that taking HRV readings after sub-maximal exercise provides similar results and reliability to morning readings,' says Manchester United's Robin Thorpe. 'It's also in our control, so that's what we do at United.'

Some, like Matt Taberner at Everton, feel the monitoring can't be controlled enough to provide worthwhile HRV results. It's the same at Paris Saint-Germain where head of performance, Martin Buchheit, won't be analysing the heart-rate intervals of Neymar and Kylian Mbappé any time soon.

'The topic of my PhD was HRV training,' Buchheit explains. 'After 10 years of trying, I felt it was impossible to implement in a football setting. You just can't have the players monitored every day, which is what I feel you need. You also really need to train hard to affect your HRV. Footballers work but not to the level of athletes in power sports like cycling, triathlon and running; it's just not sensitive to the work footballers put in. HRV finally doesn't give much information about muscular fatigue, which is important in football.'

Compress and press

Cryotherapy and HRV training are popular tools (although, as we have seen, not everyone believes in them), but they're also complicated, rather cumbersome affairs. What the players need is something more simple – and nothing comes much simpler than a pair of compression socks.

Compression socks are popular at all levels of football because of their ability to accelerate toxin clear-out. You see, every time you move your leg, your calf muscle squeezes the veins of the lower limb to send blood back up to the heart and other

A master stroke

MASSAGE IS A KEY RECOVERY STRATEGY UTILISED BY MANY PROFESSIONAL FOOTBALLERS

Research in the *Clinical Journal of Sports Medicine* in 2008 by Greek exercise physiologist Konstantinos Margonis measured changes in inflammatory and performance responses following a professional women's game of football. For example, by monitoring levels of creatine kinase, an important indicator of muscle damage, Margonis showed that it can take players' muscles up to 96 hours to recover to pre-match levels. That was 10 years ago, but his research is still cited in contemporary journals as proof of the importance of recovery in seeking peak performance over the course of a season.

A key recovery method used by most football clubs – 78 per cent of them, according to another 2008 study, this time by sports scientist David Bishop – is massage, whether by practitioner or machine. 'That looks like a traditional hot tub but it actually features jets to work on recovery and also conditions like plantar fascia issues,' said Sam Erith, Manchester City's head of sports science, when showing me one of the club's impressive training facilities. 'Those jets really help to massage the feet.'

Sports massage has many benefits, including dilating blood vessels to accelerate the removal of waste products and enhancing the speed of oxygen delivery to the muscles; relieving muscle tension and soreness; and improving range of motion – all of vital importance in preparing for 90 minutes of intense football.

A 2012 study by scientists from the Buck Institute for Research on Aging at McMaster University in Ontario, Canada, also showed that massage reduces inflammation of muscles and promotes the growth of new mitochondria, the energy-producing units in cells. Given the wealth of research backing its use, Bishop's figure from 2008 looks on the low side. Nowadays all clubs use massage in some form, whether it's administered by physiotherapists or sports massage therapists, or by the players themselves using self-massaging foam rollers.

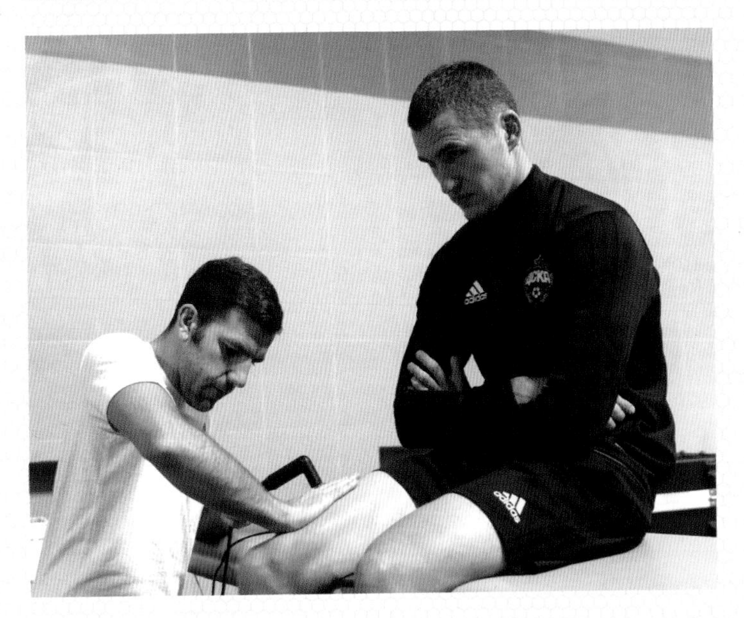

◀ CSKA centre-back Viktor Vasin receives a massage from physio Antonio de la Rubia

major organs. If you've exercised, this blood contains toxins that can inhibit recovery. These toxins are removed in the kidneys, so the more efficiently the blood can be pumped back up the body the better you will recover. Compression socks act like a second pump, speeding up this process.

Compression is measured in millimetres of mercury (mmHg), which is a unit of pressure more commonly applied to blood pressure. A normal resting blood pressure is 120/80mmHg. The first figure (systolic) is pressure of blood away from the heart; the second figure (dystolic) is pressure of blood back to the heart. Research has shown that femoral bloodflow increases to 138 per cent of the norm when the sock achieves a compression rate of 18mmHg at the ankle and 8mmHg at the calf. The higher compression at the ankle encourages the blood to flow up, not down. Of course, players' calves and ankles come in a range of thicknesses, which means that it can be hard to get to exactly those rates.

'Compression garments are big for recovery, though fit is vital,' says Robin Thorpe, head of recovery and regeneration at Manchester United. 'There's some good evidence they work, but they need to be custom-fitted to the athlete and the pressure needs to be very high. We've worked with Adidas using 3D scanners and created bespoke garments for every player. I think we're the only team in the world to do that. I can't tell you the exact mmHg figures we use. Let's just say it's a lot more than off-the-shelf garments.'

Which makes the life of Juan Mata and co. a compressed one as Thorpe asks them to wear either tights or calf sleeves when relaxing, before and after training and games, and on planes. 'But we make sure they look flash and colourful to fit in with fashion and social media,' Thorpe smiles. Some players even wear them in bed, which brings us on to arguably the greatest recovery tool: a good night's sleep…

Sleep for success

Nick Littlehales is a sleep consultant who has worked with a number of top-flight European clubs, including Manchester United, Manchester City and Real Madrid. For nearly two decades he has advised athletes on this crucial aspect of recovery.

'I fell into elite performance back in 1998,' says Littlehales, who was marketing director for bed manufacturer Slumberland at the time. 'That's when I installed the first napping room at Manchester United. Alex Ferguson wanted double-session days and didn't want the players playing computer games between sessions. He asked us what they should do. We said, let's make them sleep; let's make them nap and relax.

'Since then, I've worked with many teams, including Real Madrid. They have 80 penthouse suites at their training ground; they're like £500-a-night hotel rooms with fingerprint recognition so that only the players can get in. They're good if the players travel back late from a game.

'And then there's Manchester City, who had a blank cheque book. But that wasn't the point – we weren't trying to create a posh suite for Sergio Aguero. The work we did for them was geared around mental and physical recovery. That's why we put dawn light simulators in the rooms; full blackout blinds; took out every standby light, even on the TV; programmed messages into the TV with sleep advice and protocols; and posted information all around the training centre.'

'The players will stay here the night before if we have a lunchtime kick-off at the Etihad on a Saturday,' says Manchester City's head of performance, Sam Erith. 'In pre-season, we'll do double training sessions, so they might also use them then; it'd be training, lunch, sleep and training.'

Littlehales also worked with England at Euro 2004 and smiles when recalling the faces on the Lisbon hotel workers when the England staff arrived and started unloading sleep products from a van. 'We looked at the rooms, at things like where the sun shone when it rose and when it went down. So we put in a protocol about keeping curtains closed all the time, while also adding portable air conditioners. We then examined the bedding, which was typically European – rock-hard mattresses with crappy linen that had been endlessly washed at 60°C. The mattresses were designed to last forever and cope with 200kg – no good for a lightweight David Beckham.

'So we brought in toppers, linen and pillows, all customised to that player. For example, we put a 5cm topper on for Beckham to help him align better posturally. And it looked to be working. We were beating France 1–0 ... before Zidane scored two in injury time. Typical!'

Swansea City have had 30 sleep pods – inflatable cube-like structures – installed next to their training pitch for players to rest in between double sessions, while Bournemouth have played around with amber-coloured glasses. They might look like raver fashion glasses from the 1980s, but they're designed to reduce the effect blue light – emitted by mobile phones and laptops – has on sleep patterns. Bournemouth have conceded that they can't prevent their players trawling their smartphones in bed, so now advise them to wear these for any screen time within two hours before sleep.

And up at Celtic, head of sports science Jack Nayler says the club will hire a sleeper bus if Celtic have midweek evening games a fair distance from Glasgow. 'That said, it's still unlikely the players sleep,' he admits. Coming down after an evening match is a problem for all professional footballers. 'I worked at Madrid when Carlo Ancelotti was there and sleep was a real issue,' Nayler adds. 'In Spain, 9 o'clock kick-offs are standard; in fact, some are 10 o'clock. What with the match finishing late, there were some players who couldn't drift off until 5 o'clock in the morning.'

A study of 16 players from the German Bundesliga and Dutch Eredivisie over a 21-day period during the second half of the 2013–2014 season and first half of the 2014–2015 season highlighted the problem. The players were each asked to fill in a diary, which showed an average bedtime of 11.19 p.m. on training days, 12.18 a.m. after daytime matches and 3.27 a.m. after evening matches. Sleep duration measured, on average: 8 hours 44 minutes on training days; 8 hours 20 minutes after daytime matches; and just 5 hours 43 minutes following a night game. The players also reported feeling more restless after the evening matches. 'Suitable intervention strategies, like napping the following day, should be investigated forthwith,' recommended the study.

Performance benefits of sleep

Why is sleep so important? Let's start with human growth hormone (HGH). This hormone is released in bucket-loads from the pituitary gland during sleep. HGH repairs and rebuilds muscles by stimulating the liver and other tissues to make a protein called insulin-like growth factor 1 (IGF-1). Although players also produce HGH when they are exercising intensely, it is important that they sleep well to maximise levels and promote muscle growth. (As an aside, alcohol has long been known to stall the production of HGH. A study all the way back in 1980 in the *Journal of Clinical Endocrinology Metabolism* showed that alcohol decreased HGH secretion by 25 per cent.)

Swansea City have had 30 sleep pods – inflatable cube-like structures – installed next to their training pitch for players to rest in between double sessions, while Bournemouth have played around with amber-coloured glasses.

Disordered sleep can also affect eating habits. And, again, it's down to hormones, specifically the ones that control eating behaviour. Rising levels of a hormone called ghrelin signal that it's time to start eating, while increased levels of the hormone leptin tell you that you're full. A German study led by Petra Schuessler, published in the *Journal of Sleep Research* in 2005, showed that just one night's broken sleep significantly raises ghrelin levels, explaining why players might crave a Big Mac when they're tired. The study also revealed that two nights or more poor sleep reduces leptin levels, so you're hungry and you're less likely to recognise when you're full. In short, you can see why decreased sleep is associated with increased weight.

Sports psychologist Mitchell Smith, at the University of Technology Sydney in Australia, also studied the effects of lack of sleep on performance, and discovered that a tired brain impairs running speed, reduces passing accuracy and blunts shooting performance.

Little wonder, then, that Nick Littlehales is in such great demand to improve football clubs' sleep hygiene. 'When we worked with Nick, he brought his sleep kits,' says Southampton's director of performance science, Mo Gimpel. 'He also gave the place a clear-out as we had coffee machines and Red Bull everywhere.' Gimpel says they no longer use bespoke mattresses as transporting them for away matches was logistically challenging. 'Instead, we've just undertaken a trial with one of our sponsors, Under Armour, to test their sleep clothing. Some of the under-18s and under-23s have been trialling these kits that have a certain technology in the fabric to help sleep. We did a lot of research around it and had experts in from the University of Oxford Teensleep project.'

Under Armour's Athlete Recovery Sleepwear incorporates what's termed bioceramic technology. This is a formula that consists of over 20 types of ceramic mixed with mineral oxides, which is applied to the fabric. Ceramic materials have long been used for heat reflection, most notably on the exterior of space shuttles to reflect solar radiation. Under Armour have applied this principle to their pyjamas to absorb infrared rays emitted by the body and reflect them back as far infrared energy. Far Infrared Therapy, as it's known in naturopathic therapy circles, involves heating up tissue to increase blood circulation. The theory is that Southampton's youngsters should not only sleep better but also exhibit less soreness the day after a tough training session. Gimpel says they're still assessing the results...

APPLY THE SCIENCE 'Mineral water chaser, please'

Research supports common sense – the modern-day footballer is right to avoid the lure of alcohol

Manchester United's 1–0 home loss to West Bromwich Albion in mid-April 2018 confirmed their 'noisy neighbours' as Premier League champions for the third time in six seasons. It also gave the green light to Manchester City players Bernardo Silva, Fabian Delph, Kyle Walker and Vincent Kompany to head down to the Railway pub in Hale and lustily down pints and hand out drinks to fans who led them through the City songbook. Mid-season it would have been frowned upon by Pep Guardiola and his coaching staff. But this was a time to celebrate.

It's a far cry from days gone by and the drinking culture that pervaded English football. Take this quote to the *Daily Mail* from former Nottingham Forest striker Garry Birtles when recalling the club's 1979 League Cup Final win: 'The night before the match against Southampton we were blotto. We had everything we could possibly have wanted to drink: bitter, lager, champagne ... There were people who could hardly stand by the time we went to bed. But [manager Brian] Clough insisted on it. Archie Gemmill wanted to go to bed. He wouldn't let him. We were 1–0 down at half-time the next day, but once we sobered up, we were OK. We won 3–2.'

Clough, himself not averse to a drink or two, perhaps thought a few pints would loosen tongues, get players talking to each other and thus generate team spirit. The modern player might still enjoy the occasional drink but, in general, tales of team-bonding all-dayers are a thing of the past, as we now know much more about the negative physiological and psychological impact alcohol has on performance.

Research conducted at the Department of Clinical Physiology, at the Karolinska Institute in Sweden, showed that 'following infusion of ethanol (absolute alcohol), leg glucose uptake [of five healthy volunteers] decreased and leg bloodflow reduced, probably owing to a constriction of muscle vessels'.

In football terms, it means that if you have

As you can see from the wide range of therapies covered here, recovery is a big and important business, which is understandable when the form of your £300,000-a-week prized asset could make the difference between winning the league and finishing outside the European places. And while it's true that the evidence behind some ideas remains equivocal, this chapter shows quite how seriously the big clubs take recovery as part of their overall training programme.

Take the pioneering Milan Lab, a physiological and neurological department set up by AC Milan in 2002 to keep their players at peak fitness. It's regarded as one of the key reasons legends like Paolo Maldini, Alessandro Costacurta and current manager, Gennaro Gattuso, were able to play into their late 30s or early 40s. Of course, the fitness–fatigue tightrope is a precarious one and players still tip over into injury. But as we'll discover, the technology for managing a safe and swift return from injury is as progressive as that for optimising post-match recovery.

a few the night before a match, you will have both less energy and less oxygen being delivered to your working muscles, which will hamper recovery. Come the second half, your legs will feel much heavier than those of your cleaner-living counterparts.

Alcohol's diuretic effect has been known since at least 1948 when scientists discovered that for every gram of ethanol drunk, 10ml of excess urine was produced. Dehydration, as we will see in chapter 8, can lead to reduced performance. Alcohol's impact is even more pronounced in the heat; a study led by the wonderfully named Professor Yoda in the journal *Alcohol* in 2005 showed that alcohol impaired the body's ability to regulate body temperature.

The hops in beer contain various beneficial nutrients, such as immune-boosting antioxidants and polyphenols. Sticking to non-alcoholic beer will still give you these benefits, but without the serious drawbacks of drinking alcoholic beer. As St Mary's Clinic performance dietician Louise Davey tells me, 'After a tough football match electrolytes and muscle glycogen stores are depleted and you'll be dehydrated. Choosing a non-alcoholic beer that contains around 6–8g carbohydrates per 100ml actually unleashes the same refuelling benefits as an isotonic drink. It also strengthens your immune system with a recent study attributing its antioxidant properties to lower incidence of colds in marathon runners drinking two to three daily pints of non-alcoholic beer.'

▲ Excessive alcohol intake is in the footballing past. Now it's generally down to success like Pep Guardiola winning the Bundesliga title with Bayern

MANAGING INJURY

'This was a brown-field site; it was a dump. And fair play to the owners, they could easily have gone to leafy Cheshire. Most of the players live out there. But they said no, we want a base in the heart of east Manchester, where the fans are.' These are the words of Sam Erith, Manchester City's head of sports science, as we conclude our tour of the club's £200 million state-of-the-art training ground in the Beswick and Clayton area of Greater Manchester, funded by Sheikh Mansour and the Abu Dhabi United Group that acquired the club back in 2008. 'They spent £40 million just clearing and detoxifying the land, contaminated by the waste from old mills and paintworks,' Erith says.

When it comes to giving Gabriel Jesus, Kevin De Bruyne and co. the best, money is no object – and that includes medical provision. Negotiate the labyrinth that is the Etihad Campus and, as the sound of running water gets louder, you open a set of doors behind which is the club's hydrotherapy unit, the envy of every physiotherapist around the world.

'That's an underwater treadmill,' Erith explains. 'The floor can be raised or lowered depending on how much of the player's weight you want to carry. The deeper you go, the more weight-bearing it is. There are also underwater cameras so the player can watch their run action on the screen. It's a good tool for players returning from injury. That 20m pool over there also has an adjustable floor, which is good for deep-water running. You can also perform plyometric drills [method of increasing power through explosive actions like jumping and bounding] in there – albeit the guy in there right now is one of the support staff having a morning swim!'

◀ Manchester City's Vincent Kompany utilises sport science in search of injury prevention and peak performance

The medical area, which bears a stark resemblance to the Oregon psychiatric hospital in *One Flew Over the Cuckoo's Nest*, has space for four physiotherapists and four soft-tissue therapists. Players will be booked in, either for maintenance work or rehabilitation.

'It might be once a week, every other day or every day, depending on what the problem is,' says Erith. 'It could be something like a tight hip flexor. Things are much more professional these days. Gone are the times when players sped into the car park at 10.50 a.m. and sprinted over to training.'

Keep them fit, keep them winning

It seems obvious, but studies show that the more of its players a team can keep available for matches, the more chance that team has of winning. An 11-year study, led by Swedish physiotherapist Martin Hägglund, an Associate Professor at Linköping University, followed 24 professional clubs from nine countries and observed that 7,792 injuries were reported during over a million hours of exposure (including training and matches). On average, the 'injury burden' was 7.7 injuries per 1,000 hours, but when teams scored below this figure – in other words, had fewer injuries – they finished higher in the league. A high availability was also associated with success in the Champions League and Europa League.

Hägglund's conclusions were borne out over two recent Premier League seasons. In the 2015–2016 season Leicester City players were out of action for a mere 192 days,

▼ Leicester players were out of action for a record low 192 days on their way to the 2015/2016 title

> Things are much more professional these days. Gone are the times when players sped into the car park at 10.50 a.m. and sprinted over to training.
>
> **SAM ERITH** HEAD OF SPORTS SCIENCE, MANCHESTER CITY

the lowest in the league and 81 per cent lower than the league average of 1,035.5 days. Leicester's figure rose to 885 days in 2016–2017 – admittedly still below average – as the Midlands side briefly flirted with relegation before recovering to finish 12th.

In 2016–2017, after an indifferent start that culminated with a 3–0 loss to Arsenal, Chelsea boss Antonio Conte reverted to his more familiar three-man defence and the team proceeded to win 13 consecutive games for the first time in club history. Their title-winning final tally of 93 points was the third highest ever in the Premier League.

Many identified that three-man defence as the foundation for silverware and clearly it had an impact. But it also can't be ignored that Chelsea lost a collective 877 days to injury – significantly above Leicester's freakishly low 2015–2016 number but still the third lowest in the league. Importantly, the London club also recorded the fewest injuries lasting over 14 days that season – only 12 – compared to 39 at Sunderland, who finished bottom.

Tugging at the hamstrings

However you look at the figures, clearly a lot of days are lost to injury. So what are the most vulnerable areas? 'The hamstring is the biggest problem,' says Everton's strength and conditioning coach, Matt Taberner, from the club's Finch Farm training ground. 'Then it's knee, ankle and groin issues.' At the time of my visit, Taberner and his team were tending to Phil Jagielka and Ross Barkley, who both had hamstring problems, as well as long-term absentees Seamus Coleman (broken leg) and Yannick Bolasie (anterior cruciate ligament injury). Taberner's experience matches the data (see box).

Top five reasons for missing training/match days

This table shows the five reasons that caused the most Premier League players to miss training and match days during the 2015–2016 season.

	PLAYERS AFFECTED	TRAINING/MATCH DAYS LOST
Hamstring	150	4,165
Knee	119	3,396
Ankle	101	2,650
Illness	72	459
Groin	60	1,336

Tales of hamstring woe continued to dominate the back pages going into the 2017–2018 season, with world-class players like Paul Pogba, Sadio Mané and N'Golo Kanté all missing at least a month. This represents a growing and alarming trend in professional football, despite clubs having access to cutting-edge medical provision that puts the old magic sponge to shame.

▲ Paul Pogba suffers a hamstring injury in the 2017 Champions League group game win over Basel

Professor Jan Ekstrand is an orthopaedic surgeon and vice-chairman of the UEFA medical committee. In 2016, the Swede and his team published a study that showed hamstring injuries in men's professional football had increased annually by 4 per cent between 2001 and 2014. Anterior cruciate ligament (ACL) injuries had also increased by, in one study, 6 per cent a year. That compared to groin injuries, which have seen a yearly 2 per cent decrease since 2001, and medial ligament injuries, which have gone down by nearly 7 per cent.

Interestingly, that increase in hamstring injuries stems mainly from training incidents (although Pogba, Mané and Kanté all sustained their recent injuries during matches). Ekstrand suggested that this might be because 'the focus of training sessions included more repeated high-intensity actions that replicate the evolving nature of the game.' This is supported by his observation that '70 per cent of hamstring injuries occurred during sprinting or high-speed running'. That certainly fits with the picture we drew in chapter 1 of increasingly intense Premier League training sessions.

At this rate, UEFA predicts that 20 per cent of all football injuries will be hamstring-related by 2032, which would be double 2001 levels. So what is going on? Glen Driscoll is performance director at Glasgow Celtic. He joined in May 2016, hooking up with manager Brendan Rodgers for the third time after stints together at Swansea and Liverpool. Driscoll also spent nine years at Chelsea, dealing with areas of fitness and conditioning, physiotherapy and injury prevention.

After leaving Liverpool, Driscoll had some time off. Too much time. He was bored. So he did what any scientist worth his salt would do and spent hours wading through data. 'I noticed that at Liverpool, whenever we played a pretty intense full-pitch tactical game the day before match day, soft-tissue injuries, especially hamstring problems, increased,' he explains. 'Why is it that high-intensity work, in space, across the most acute build-up phase of pre-match training caused such issues?'

Driscoll found the answer in research by US biomechanical expert Elizabeth Chumanov. In 2007, Chumanov wrote a study entitled, 'The effect of speed and influence of individual muscles on hamstring mechanics during sprinting'. She analysed the running gaits of 19 athletes sprinting on a treadmill with speeds ranging from 80 to 100 per cent of maximum velocity. 'Chumanov showed that we reach full stride length at 80 per cent of full velocity, which is after about 10m of sprinting,' explains Driscoll. 'Moving from 80 to 100 per cent of full velocity, there's an exponential increase in hamstring force, which is accompanied by negative work of the hamstrings. Put another way, at full stride length there's a significantly increased strain on the hamstring that predisposes that muscle to injury.'

It's why there are very few hamstring injuries in tennis. While intensities are similar to football, a tennis court is much smaller (23.77m long by 8.23m wide) than a football pitch (the Premier League's ideal is 105m long by 68m wide). If a tennis player hits his or her full cadence, they'll either end up tearing through the net or toppling into the crowd. 'It's why small-sided practice games on smaller pitches are a good way to reduce hamstring injuries,' Driscoll continues. 'There simply isn't the space to reach full stride.'

The problem is, while small-sided games increase intensity and demand a faster speed of thought, they don't fully prepare the player for match day. That's where flexibility and innovation comes in. 'Our GPS data shows that if the session involves small areas, the player won't be able to reach highest speed because there simply isn't the space,' explains Martin Buchheit, head of performance at Paris Saint-Germain. 'Your biceps femoris [one of the four hamstring muscles] is simply not lengthened. That's why we'd often prescribe specific hamstring work in the gym after a small-sided session. Conversely, if there's a lot of space in a session, we know the players will be touching high speeds, so we'll lay off the gym hamstring work.'

> At full stride length there's a significantly increased strain on the hamstring that predisposes that muscle to injury.
>
> **GLEN DRISCOLL** PERFORMANCE DIRECTOR, GLASGOW CELTIC

Measuring eccentric muscle strength

Buchheit remains tight-lipped about the specific hamstring exercises he uses. Others, like Dave Tenney, former head of sports science at Seattle Sounders, are more open. 'We use the NordBord,' he says. 'Essentially it's a machine where you undertake Nordic curls and it measures the eccentric strength of each leg.'

Nordic hamstring curls are where you clamp both feet into straps, while kneeling up straight on the floor, and slowly lowering your body to the floor before rising up again. They're designed to increase eccentric muscle strength (where you increase tension on a muscle as it lengthens), because a lack of eccentric strength is one of the main reasons behind hamstring injuries.

Killing the pain ... at what cost?

PAINKILLER (AB)USE IS REPORTEDLY AT ENDEMIC LEVELS IN PROFESSIONAL FOOTBALL WITH LIFE-CHANGING CONSEQUENCES

'I've been in many dressing rooms where I've seen players pressured into playing, myself included. I had painkilling injections for a broken toe for six months – one before the game, one at half-time ... I'd wake up at midnight in agony as it wore off with a very sore big toe. It was my choice. Was it good for me long-term? Arguably not. Could it lead to arthritis? Possibly.' The words of former England defender Danny Mills in a 2017 interview with the BBC.

Mills isn't alone in playing through pain. According to FIFA's former chief medical officer, Jiri Dvorak, 60 per cent of players at the 2010 World Cup were administered some form of painkiller at least once during the tournament. These included non-steroidal anti-inflammatory drugs (NSAIDs) like ibuprofen, local anaesthetic injections and corticosteroidal injections. The data was collected from lists provided by team doctors of medications used by each player within the 72 hours before each match.

It highlights the very real issue of clubs and international teams putting success over player welfare. You might think that over-the-counter drugs like ibuprofen are harmless. 'They're not,' says Dvorak. 'You can't take them like cookies. They have side effects.'

Overuse of these drugs can be detrimental in two ways. The first is that a player trains or competes when they shouldn't, exacerbating the injury. In a 2011 interview for online magazine *Sabotage Times*, former Liverpool and Leeds defender Dominic Matteo spoke of regular painkilling injections at both clubs that masked back issues, which ultimately resulted in spinal surgery and retirement from the game.

Then there are the issues created by the medication itself. The National Institute of Health (NIH) in America, one of the world's foremost medical research centres, warns, 'NSAIDs such as ibuprofen may cause ulcers, bleeding or holes in the stomach or intestine. These problems may develop at any time during treatment, may happen without warning symptoms and may cause death.' A further study by the NIH concluded that NSAIDs weaken the intestinal walls, and that long-term use of these drugs leads to inflammation of the small intestine.

In 2016, Croatian striker Ivan Klasnic won a legal case against his Werder Bremen doctors after successfully arguing that their failure to properly diagnose his kidney problem – and the continual prescription of painkillers – led to a severe deterioration in his condition. In 2017 Klasnic had a third kidney transplant after the first two failed.

The solution – a more rigorous independent assessment of players – doesn't appear forthcoming as Professor Dvorak previously raised these concerns when he was employed by FIFA (he left in November 2017), but claims the world governing body has still not addressed the issue appropriately.

▶ Nordic hamstring curls are a key exercise in the battle to stay fit

In a number of studies across Danish and Dutch elite and amateur players, hamstring injuries decreased when following a Nordic hamstring programme. Take the 2015 study by hamstring expert Nick van der Host, which had 292 Dutch amateur players performing 25 Nordic hamstring sessions over a 13-week period compared to the control group of 287 amateur players who performed no hamstring exercises. While the Nordic group had just six hamstring injuries, the control group suffered three times that number at 18. It's why Nordic hamstring curls are a staple at every club with an increasing number using the NordBord Hamstring Testing System.

'The NordBord is great because it sends information to a data-collection app [ScoreBord] that helps you to paint a picture of how robust a player is,' says Tenney. 'You see, each player has their own threshold. [From testing using the NordBord we can see that] one player might generate 250 Newtons (N) of hamstring force before they're raising their potential of injury; whereas for another it might be 400N. As well as preventing injury and improving performance, the NordBord is also useful for guys returning from injury – especially ACL – because it shows you how far a player is from baseline.'

The system is currently used by 16 of the 20 Premier League clubs and, says Everton's Matt Taberner, Nordic hamstring curls are a popular component of many injury prevention and conditioning programmes. 'However, many players who don't possess a sufficient strength base fail to execute the correct technique and full range of motion. That's why I'd recommend the sliding leg curl,' he says. Crudely put, this is where the player lies on their back with knees bent and feet flat on the floor, pushes their pelvis up into the air and slides their feet back and forth along the floor, keeping their bottom off the ground throughout. It remains an eccentric exercise, so stresses the hamstring as intended, but is technically easier.

Protecting the tactics

Medical support staff are clearly utilising the latest research and technology to slow the alarming rise in hamstring injuries. But equally important are their experience and the clarity of their communication channels to the manager. Over to Southampton's first-team physio, Steve Wright, who's been in the game for over 15 years.

'How a manager sets out his team tactically can predispose players to a particular injury,' he says. 'When [Mauricio] Pochettino was here, to activate his pressing style he trained every single day at a really high intensity, which predisposed the players to hamstring injuries. So we did a lot of hamstring-injury-prevention work. But we didn't need to do the same with [Ronald] Koeman as he played a different style of football involving less pressing.'

Wright says that while hamstring injuries dropped in recent seasons, knee injuries went up. So the team would go through what they call an 'analyse-prescribe' routine, beginning with observation and biomechanical assessments. 'Say the anterior cruciate ligament (ACL) goes when the knee buckles on the inside,' he says. 'That could be down to landing mechanics – but is that a strength issue or a neurological one? Once

▼ Mauricio Pochettino's pressing style meant the likes of Luke Shaw (Southampton, 2014) required significant hamstring-strengthening work

you've identified the problem, you can then put a strategy in place. In this case, if a player looks vulnerable to ACL we might build strength around the hip, because if the knee collapses it's often the hip that brings it back. We'd do a lot of landing practice on soft mats, with occasional work on a harder surface so that it's relevant to match days.'

Wright then gives me a wry grin as he concedes knee injuries might have dropped off in 2016–2017 but hamstring injuries went up again – so much so that the club brought in Dr Katie Small from the University of Cumbria, an expert in hamstring injuries. Small told website Training Ground Guru: 'The number of hamstring injuries Southampton had last season was pretty standard – around 11.5 per cent of their total injuries. It was more the severity of them that was a concern. Rather than players being out with a hamstring injury for 21 to 28 days, which is the average, some of them were out for more than 100 or even 200 days.'

Small's work with Southampton is ongoing, but if her previous research is anything to go by she might focus on the timing of hamstring-strengthening exercises within training sessions. Her 2009 PhD work identified the ends of each half of a match as the times when a disproportionately high number of hamstring injuries occur. With this in mind, she sought a training strategy that would maintain eccentric hamstring strength throughout a match. During an eight-week trial involving two groups of semi-professional players, she found that the players who did hamstring-strengthening exercises right at the end of a training session, when the muscle was fatigued, maintained more eccentric strength during a simulated match than the players who did the exercises at the beginning of the session.

Managing congestion

Something that medical staff can't do anything about is the very thing that many insiders argue is the main reason for injuries – the packed fixture list.

Manchester United manager José Mourinho spent much of spring 2017 bemoaning fixture congestion. This was a period when the Red Devils were becoming victims of their own success. During the run-up to the Europa League Final, in which they beat Ajax 2–0, they played four games in 10 days, en route to a season total of 64 games. That beat the club record 63 games they played during the treble-winning 1998–1999 season. 'We have two cruciate ligaments ... and the other small injuries, the Pogba one and the Valencia one, they are injuries of fatigue,' Mourinho complained after a midweek Manchester derby on 28 April.

He's certainly supported by a number of studies, including a 2017 paper by Jan Ekstrand published in the *British Journal of Sports Medicine*, that showed a much greater risk of muscle injury when matches are separated by two to five days than when matches are separated by six days or more.

Unlike Manchester United, Chelsea weren't in Europe in 2016–2017 and racked up just 47 games. Did increased rest and recovery, and no draining travel, result in a fresher, fitter squad? 'It's difficult to compare because nothing ever stays the

▲ Fixture congestion didn't stop Man United winning the 2017 Europa League Final

same,' says Chelsea first-team physiotherapist, Jon Fearn. 'When a manager changes, training changes. With previous managers, we didn't do any gym work, for example. The current manager [Antonio Conte at the time] and his team do gym work. Things change. We don't tend to do weights now, either; it's very functional. It's a lot to do with improving balance and proprioception [awareness of body position and movement].'

Fearn also says that having fewer games allows the manager and his team to push the players harder in training. 'There are many variables,' he continues, 'but for me the most important thing is the length of injury. Yes, the squad might have had 10 hamstring injuries, but each player might only have been out for a day.'

Learning from experience

Injury prevention relies heavily on a combination of data collection and experience. Until recently, Tony Strudwick was head of performance at Manchester United and had been with the club for more than 10 years. 'Sir Alex [Ferguson] would carry a big squad of 26 players,' he says. 'But our sports-science model revolved around a few different objectives. One was to extend the careers of Ryan Giggs, Paul Scholes and Rio Ferdinand, so they needed micro-management. Another was to bring in the younger players like Danny Welbeck, Phil Jones and Chris Smalling.

'Would you believe, in his last 250 games, Sir Alex only picked the same team in consecutive matches two or three times,' he continues. 'He'd often make four or five rotations a game. He always kept the plates spinning. We created player-performance metrics so we had objective data to marry with Sir Alex's intuition.

'Take Ryan Giggs. We were recording things like how many minutes he played, how many training sessions, looking at wellness and recovery states. We did that for many years, just building up a pattern. It was a bit like training a thoroughbred horse – you come to understand what they can and can't do. From the information we gathered, we knew that he was best when playing a game every 11–12 days. Anything more, it could lead to injury; anything less, he'd lack match sharpness.'

Clearly, United handled Giggs well. The Welshman racked up a club record number of appearances and won BBC Sports Personality of the Year in 2009 at the grand old age of 36. Strudwick also makes the point that some players are just naturally more robust than others – apart from that metatarsal, for example, how long did David Beckham spend on the sidelines? How many times has Cristiano Ronaldo been injured?

It's no coincidence that a club's most prized players are also often the most resilient, both physically and mentally. As Sweden's Martin Hägglund showed earlier, keep them on the pitch and the results will come. But that's no easy job in the Premier League. Even clubs lacking the success of Manchester United face yearly winter congestion.

▼ 'More than a game every 11–12 days would have increased the chance of Giggs being injured,' says Tony Strudwick

'Tis the season to get injured

During late December and early January, English clubs can often play four games in fewer than 12 days – a situation that left former national coach Roy Hodgson lamenting, 'I don't believe in the English footballing calendar. I'm only talking from a technical point of view what I as a coach or a player would want. Not talking about money, TV rights, sponsorship … everything I say now could be naïve. But I think it's unbelievable that in England we have teams who finish playing on, say, 6 May and won't play another competitive game until 15 August, and then from August to May they will play 55 games. I don't understand that.

'With football, it's not necessarily a physical fatigue thing – it's mental, too. Getting yourself up for a game for so long. Saturday, Tuesday, Saturday … up for the level of performance required. You just need to leave football behind for a while [during the season] and forget that it exists.'

Hodgson recommended a three-week winter break and a shorter off season, but didn't think it would ever happen. Many argue that it is little wonder Germany, whose Bundesliga players enjoy a three- or four-week break each winter, are consistently successful in World Cups and European Championships. And they're not the only league to give players a mid-season rest: France had 24 days off over Christmas last year, Spain 18 days and Italy 16. Premier League clubs had a maximum of three days off during the festive period.

> With football, it's not necessarily a physical fatigue thing – it's mental, too.
>
> **ROY HODGSON**

While foreign league players enjoy some Dubai sun and a brief second off season, in England they're racking up injuries. Figures for the 2017–2018 Premier League season show that injuries rose to a high of 143 players in January 2018 as the congested Christmas fixture list took its toll (see box).

Despite Hodgson's understandable pessimism, things are set to change. In June 2018, following discussions between the Football League, Premier League and Football Association, it was announced that Premier League clubs will have a 10-day winter break from February 2020. (This is achieved by moving the FA Cup fifth round matches to midweek and splitting one round of Premier League matches between two weekends; half the teams will break slightly earlier than the other half.) It doesn't match Germany, but it will be more recuperative than the current scheduling.

Seasonal injury trends

THE NUMBER OF INJURIES IN THE PREMIER LEAGUE FLUCTUATED CONSIDERABLY DURING THE 2017–2018 SEASON

With data from the invaluable website physioroom.com, here's how injuries affected Premier League clubs month by month during the 2017–2018 season correlated to matches played (league matches only). As the season cranks up around the Christmas period so do the injuries. Things settle down in February before accumulated fatigue takes its toll in March. The information is from 1 August 2017 to 1 May 2018.

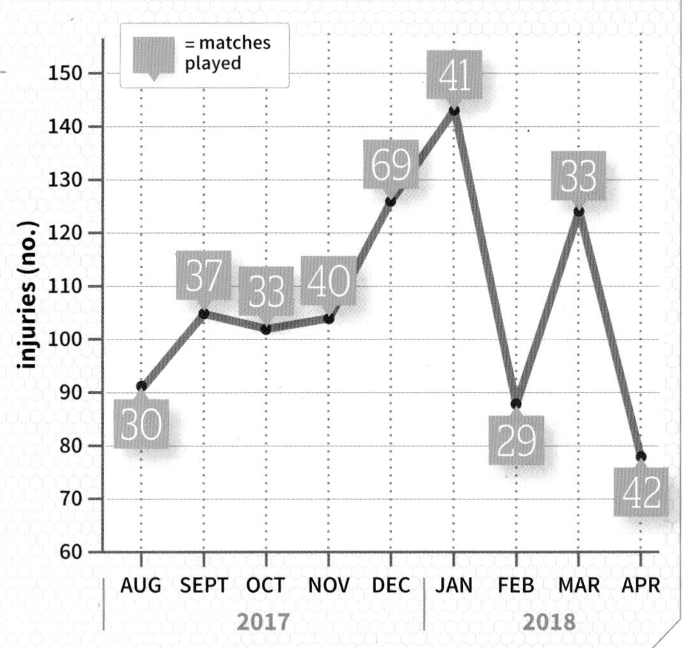

▶ Charlie Austin suffers a dislocated shoulder against Hapoel Be'er Sheva in December 2016

Case study: **Charlie Austin**

Winter break or not, players will always get injured and need a support team around them to accelerate recovery.

In December 2016 one of the 113 Premier League footballers on the treatment table (according to statistics for missed training days published on the website physioroom.com) was Southampton striker Charlie Austin. Austin, a £4 million signing from QPR the previous January, had scored six of the team's 13 Premier League goals in the first part of the 2016–2017 season before dislocating his shoulder in the Saints' 1–1 Europa League draw with Israeli side Hapoel Be'er Sheva on 8 December 2016. First-team physio Steve Wright was charged with returning Austin to duty as quickly as possible, but without cutting corners that could lead to a recurrence. Here, Wright describes in detail the hurdles Austin and the medical team faced – and eventually overcame – on that torturous path.

'Charlie has dislocated his shoulder three other times in his career – two on one side, one on the other – and had previous procedures done at other clubs before coming to us. He'd seen the best guy in the UK. It went well but, from a psychological standpoint, we had to do something different to give Charlie the belief that he'd come back even stronger than last time. So we went to the Steadman Clinic in America, which made its name in knee surgery. It happens to be in the snowy mountains of Colorado where they get a lot of trauma injuries from skiing, so they have plenty of experience.

'Back in the day we used to find the best surgeon near to us. Now we look worldwide. He had surgery within a week of the incident. Fair play to the club, this wasn't an insurance job – there won't be much change from $100,000 for surgery and two appointments.

'I went over with Charlie and watched the operation. He was lying on his side with his arm in the air for an hour under general anaesthetic. They went in through the back of the shoulder – a six-inch incision. It was useful to watch to see which muscles they cut through because that will need specific rehab. Scans are informative but they're always building up a general picture rather than giving you the specific view you get with a camera. I could see that the whole shoulder would require rehab work after six to eight weeks in a sling, but a couple of muscles at the back of the shoulder would be even weaker because they were divided and sewn back together.

'When we could, we did some mobility work. In terms of fitness, we did lots of low-impact stuff like using a cross trainer. Because Charlie has some knee issues as well, we didn't want to do too much cycling. We did a lot of deep-water running, too, where he wore a flotation device.

'We also used an altitude device to create a degree of hypoxia [oxygen deficiency] to make his system work that bit harder; and we got him up to 12,000ft oxygen-wise. We do this with most of our injured players because predominantly we're dealing with lower-limb injuries where you're limited in the load you can apply. Hypoxia compensates for this lack of muscular loading by increasing the cardiovascular load.

'Another piece of equipment we used was an anti-gravity treadmill. The player wears what looks like a ballet tutu and then steps into an airtight waist-high chamber on a treadmill. The chamber is filled with air that supports their body; it can take up to 90 per cent of the body's load. (In fact, Virgil [van Dijk, who has since joined Liverpool] is on it at the moment, even though he's still on crutches with a foot injury. Virgil's playing weight is 99kg. We can offload him to make him weigh 20kg.) We go to great lengths; in fact, I also went on holiday with Charlie to make sure he remained committed to the physio programme we'd prescribed for him!

'We're four months down the line now. Last week we went back to the US and got sign-off for higher-impact work.

'Now it'll be me, the sports scientist and a rugby crash pad. We want Charlie to have confidence in his shoulder. We can measure strength and mobility but psychologically we need to get rid of any demons. He uses his arms a lot to play; he gains leverage off other players to go in the air.

'We'll then drip-feed him into training over the next couple of weeks. Of course, the manager and coach are going to lead the sessions, but if they're happy for us to do so – if it's four or five days out from a game, for example – we might be able to impose some rules on the drill. For example, we can put Charlie in as a floater, so he plays for both sides. That means he receives the ball and passes it but no one's going to tackle him. We've also done shooting drills for the last six to eight weeks.'

Fuelling rapid recovery

CAREFUL ATTENTION TO A PLAYER'S DIET CAN RESULT IN A QUICKER RETURN FROM INJURY

The diet chart below was published in a 2014 paper in the *International Journal of Sport Nutrition and Exercise Metabolism* called (deep breath): 'Case study: Muscle atrophy and hypertrophy in a Premier League soccer player during rehabilitation from ACL injury'. This explains how a team led by Liverpool's strength and rehab fitness coach, Jordan Milson, maintained muscle mass and function of a player recovering from an anterior cruciate ligament injury.

Milsom and his team carefully planned the injured player's diet to increase protein intake during the first eight weeks while reducing carbohydrate intake. This limited the loss of muscle mass during the period of immobilisation, while keeping carb calorie count down, which would normally be needed to fuel intense training sessions and matches.

During the rehabilitation phase, as the player became more active – first in the gym and then out on the field – carbohydrate intake increased to match the energy demands. Protein intake remained high, again to maintain and then increase muscle mass.

The player also took a supplement of HMB (beta-hydroxy-beta-methylbutyrate) – an amino acid that's been linked to preserving muscle tissue – and creatine. We expand on creatine pros and cons in our examination of match-day nutrition in chapter 6.

MEAL TIME	IMMOBILISATION PHASE	REHABILITATION PHASE
Breakfast	Three-egg omelette including ham, tomato and cheese; 400ml semi-skimmed milk; one multivitamin, 500mg vitamin C, 1g fish oil and 1.5g HMB (protein supplement)	One bowl of muesli and 250ml semi-skimmed milk; two slices of wholemeal toast and three poached eggs; 350ml apple juice; one multivitamin, 1g fish oil and 1.5g HMB
10 a.m.	30g whey/casein protein supplement; 5g creatine monohydrate	25g whey/casein protein supplement; 5g creatine monohydrate
1 p.m.	Two chicken fajita wraps including salsa, peppers, onions and mushrooms; mixed salad; 1g fish oil	Chicken fillet, sweet potatoes and mixed vegetables; 350ml apple juice; natural yoghurt and mixed berries; 1g fish oil
4 p.m.	30g whey/casein protein supplement; 5g creatine monohydrate	25g whey protein; 50g carbohydrate recovery shake; 50g carbohydrate energy bar
7 p.m.	Salmon fillet, basmati rice and broccoli; 1g fish oil	Fillet steak and mixed salad plus potato wedges; 350ml apple juice; 1g fish oil
Around 30–60mins before sleep	30g whey/casein protein supplement; 1.5g HMB	30g whey/casein protein supplement; 1.5g HMB
NUTRITIONAL BREAKDOWN	1,970 calories: 140g carbs, 195g protein and 70g fat	3,170 calories: 400g carbs, 190g protein and 90g fat

Austin made his return to the first team on 13 May 2017, 111 days after dislocating his shoulder. It helped. Southampton won their penultimate game of the season 2–1 away to Middlesbrough. Austin scored a late penalty on 19 August 2017, his first goal since the injury. He went on to score five more goals before succumbing to – yes, you guessed it – a hamstring injury just before Christmas, which kept him out for another 98 days.

Case study: **Jack Wilshere**

Jack Wilshere is generally regarded as one of the most talented English midfielders of his generation. He came up through Arsenal's academy, becoming the club's youngest-ever debutant in 2008 at just 16 years and 256 days. At 18 years and 222 days he became the 12th-youngest player to win a full England cap.

All looked rosy as Wilshere established his place in the team, racking up 35 Premier League appearances in season 2010–2011 … but that's when his problems started. During a pre-season match against New York Red Bulls in July 2011, he suffered a stress fracture of the ankle that ended his season before it had begun. A litany of injuries over the next seven seasons, including a hairline crack in his calf bone and a knee operation, meant he would average just 16 Premier League games a season during this period before signing for West Ham in June 2018.

◀ Despite the best efforts of medical staff, Jack Wilshere's (right) career stalled through injury

Boss Arsène Wenger was a great supporter of Wilshere, but by the end of the 2015–2016 season – of which he missed all but three matches – everyone realised that the midfielder wasn't going to be able to get the game time he needed in the Arsenal first team. So it was decided that he should go out on loan.

'The chief executive came into the office and said, "Don't tell anyone, but we're on the verge of signing the highest-profile player this team's ever signed,"' Craig Roberts, the Bournemouth club doctor, recalls. 'So I did what any good doctor would do and googled his injury history!'

> Once we pinned down his [Wilshere's] game metrics, we got an idea of his match-day load and worked back from there to work out his weekly load.
>
> **CRAIG ROBERTS** CLUB DOCTOR BOURNEMOUTH

Roberts spoke to the medical team at Arsenal to find out in detail about the overuse injuries, the tendon problems on both sides and the stress fractures... 'It was a pretty tough picture for a really good player.'

It was clear Wilshere couldn't handle the load of three games a week. It was also clear that this was one special player who could make a big difference for Bournemouth. 'Before we made a decision, we got the whole medical team together and asked, do we think we can get him through a season? Yes or no? We needed experts in different fields and that was key. Medical, sports science, manager, assistant manager ... we needed their buy-in.

'Sometimes the player can feel overloaded with detail when you put a team of this size together to work on their rehabilitation. So we adopted a "modified athlete-centred" approach. You sit around a table once a month and have a scheduled meeting, with the player chairing the meeting, so that he drives the process and we are the experts around him. It's a subtle point but important because the player feels he's in control of the process.'

Roberts and his team collected all the data from Arsenal on Wilshere. 'Because he'd played so little in the previous season, we realised that training load was going to be critical, particularly in the first three months,' says Roberts. 'We looked at his high-speed runs and distance covered. We were aware of ground contact time and stress reaction from that so we stayed away from high-speed running initially and substituted that with high-speed exercise off the pitch. We'd do gym work or hypoxic sessions on the exercise bike. One of the things the manager bought into was our assertion that he wasn't ready to play a full match from the start. In fact, it took him eight weeks to reach 90 minutes.

'Once we pinned down his game metrics (distance run, sprints, accelerations...) and he grew stronger, we got an idea of what his match-day load was and worked back from there to work out his weekly load. Generally, match day -1 was a light session, tactical and covering about 3.5km in total. Match day -2 was a day off from being on the field but a hard conditioning day in the gym. His hardest training days were match day -3 and match day -4. We knew how much he could do and had a target. I wasn't the most popular person because as soon as he hit the GPS target we'd set for him, I'd walk on the pitch and take him off. Jack hated it. But it slowly built up his chronic load.'

Roberts worked closely with manager Eddie Howe. Often, they'd agree that Howe could have him for two sessions a week, while the medical team would work on him the other two sessions. 'But, and this is important, he'd always warm up with the other players. That camaraderie is vital.'

Wilshere's short, sharp sessions gradually increased, but the heavy strength work in the gym continued. This, Roberts says, was also vital. 'There's a misconception in football sometimes that too much gym work equals bulk. That's wrong. It's vital to build up strength to prevent fatigue and the player breaking down, so we did a lot of gym training with Jack, particularly on areas like his tendons where he's struggled in the past.'

And how did Wilshere respond? 'We were quite happy,' says Roberts. 'We built him up slowly and, over that period, he played in 27 of the 30 Premier League games and took part in 97 per cent of the training sessions.' So all good? Nearly. 'Just when we thought we were on top of things, with five games to go he suffered a crack in his fibula away to Spurs and that was season over.'

As this case study highlights, strong internal communication is one of the key factors in helping a player to return from injury. Managerial leadership was the focus of another study by Professor Jan Ekstrand in 2017, who asked if there was a correlation between coaches' leadership styles and injuries in football.

Ekstrand and his team questioned medical staff at 36 elite football clubs in 17 European countries (the clubs remained anonymous). As well as gathering information on injury rates and player availability, the researchers elicited the medics' opinion of the club's manager by asking them to answer the seven-point Global Transformational Leadership questionnaire. Based on the medics' answers, the manager's style of leadership was assessed as 'transformational' (democratic, participatory), 'transactional' (authoritarian, directive) or 'laissez-faire' (lacking in clear direction). Ekstrand concluded that 'teams whose managers tend to employ a democratic leadership style have a lower incidence of severe injuries.' In fact, the severe-injury rate was 29–40 per cent lower in clubs where coaches 'communicated a clear and positive vision of the future, supported staff members and gave staff encouragement and recognition'. It might all sound a touch corporate, but this injury-reducing clarity

There's a misconception in football sometimes that too much gym work equals bulk. That's wrong. It's vital to build up strength to prevent fatigue and the player breaking down.

CRAIG ROBERTS CLUB DOCTOR, BOURNEMOUTH

▲ Former Aston Villa striker John Carew enjoyed the unenviable record of two failed medicals

of communication also had an impact on the training pitch, where attendance was reported as higher in clubs where coaches exhibited that democratic style.

Without knowing the internal machinations of individual clubs, it's hard to gauge which managers sit beneath each leadership umbrella, but we can all hazard a guess based on their conduct in the media. Whether these results will influence clubs' managerial selection process remains to be seen.

Passing the medical

Injuries strain a club's finances as well as its players' muscles and ligaments. In 2016–2017, according to physioroom.com, the Premier League injury wage bill reached £180,437,380 – an average of £9,021,869 per club. That's why the medical that all players undergo before signing at a new club is no mere formality. Speaking at the 2017 Future of Football Medicine conference in Barcelona, former Blackburn Rovers head of sports medicine Dave Fevre described the process.

'Our job as a medical team is, in essence, to be detectives, to find out if anything's wrong with a player coming into the club. You wouldn't buy a car without it being tested by a mechanic and it's the same with a player. Why buy a player for £10 million when you can knock that price down to £8 million or £6 million? Or if we can get a player on a two-year deal instead of three years, that might save the club a million pounds. Say we found this thing on the player, we're worried about it and need to drop contract and salary. That's part of my job.'

When it comes to the 'medical' involved in transfer dealings, Fevre says there's no standard procedure, telling *FourFourTwo* magazine, 'There are key elements that most medicals will feature, though. A health check to look for any cardio concerns is one of these; since Fabrice Muamba's cardiac arrest, clubs are even more diligent about checking any heart irregularities.

'You'll look to sign players who have consistently played, and it is then up to you to provide the manager with an assessment of where they are physically and what additional work may be needed to get them up to full fitness. The club will use their own checks for this, everything from body fat percentages to VO_2 max testing to gauge where the player is at and how far away they are from starting.

'Some clubs will include vision, hearing and dental checks during the medical. For example, during my time at Manchester United we worked with Professor Gail Stephenson [vision scientist at Liverpool University] in looking at many aspects of sport, exercise and vision.'

▼ Cardiac surgery and a dip in form confined Kanu's Inter career to 12 games

Ultrasound or MRI scans are common if the player has had a long history of injury, and, sometimes, he fails. Former Norwegian striker John Carew failed two medicals during his career, once for Fulham in 2002 and then 11 years later at Inter Milan, while French left-back Aly Cissokho failed a medical with AC Milan back in 2009 because of dental issues. Speaking to the *Sun* when playing for Aston Villa in 2014, Cissokho said, 'Milan reported that there was some kind of problem with my body, which was linked to my teeth! But it was just an excuse because I completed the medical in two hours with no problems!' Medical expert and Italian club Ascoli's medic Renzo Mandozzi later told Goal.com that it could have been linked to the presence of granulomas (inflammation of the tissues associated with tooth rot), which have been noted to affect human muscle or the heart.

Roy Hodgson has been managing for over 40 years. His stock rose in the mid-1990s when

APPLY THE SCIENCE

From RICE to POLICE

When it comes to treating injuries, the boys in blue have won the battle of the acronyms

A crunching tackle. A victim writhing in pain. And this time the pain is genuine. 'That's where you adopt the old acronym RICE – rest, ice, compression, elevation,' we hear you cry. Not any more. A 2012 study by Chris Bleakley, Philip Glasgow and Domnhall MacAuley, featured in the *British Journal of Medicine*, proposed it was time to call the police.

'For acute injuries, we use POLICE,' explains Tom Goom, head physio at Brighton's Physio Rooms. 'P is for protect, so strap it up; O and L stand for optimal loading. They're the major changes. With RICE you were advised to rest, but it's been proven to reduce recovery time if you can keep a certain amount of movement. Of course, some people will be on crutches, but many will still be able to apply some form of pressure. Ultimately, just stay within comfortable pain limits.'

As for the remainder of the acronym, that's the same as before – ice, compression and elevation. Those three measures are designed to reduce swelling, which is a big factor in acute problems like ankle sprains. It's common wisdom that you need to reduce swelling, but why? Surely it's the body's instant method of relief?

'That's a good point. Yes, swelling is part of the rehabilitation process and it brings healing chemicals to the area of the injury,' replies Goom. 'The downside is that swelling reduces activity around the joint, so it leads to muscle weakness and instability. It also tends to make the joint painful and stiff. So if we can reduce that, you'll probably get movement and feeling back much quicker.'

he lifted the Swiss national side to third in the world rankings. Serie A giants Inter Milan took note and duly appointed him as their head coach. Hodgson spent three seasons at the San Siro, taking Inter to the final of the UEFA Cup in 1997 before losing to German side Schalke 04.

It was while managing Inter, in 1996, that Hodgson fully understood the importance of rigorous medical screening. 'We bought [Nwankwo] Kanu off Ajax,' Hodgson told the audience at the 2017 Future of Football Medicine conference. 'He'd won a Champions League medal with them the year before. When we bought him, he went straight off to the Olympic Games, where he captained Nigeria to gold.

'When he returned to Italy, we had little preparation time. But he played his first friendly and scored, and played even better in the second game, scoring two goals. We thought we've cracked it here, as we needed a good centre-forward.

'Then he went for a cardiac screening and came back diagnosed with a heart condition and, apparently, couldn't play football anymore. The doctor said, "I'm giving you medical facts. I'm saying to you he might play for another 10 years and not have a single problem. But he could also be walking down the street and drop down dead any minute. So I'm telling you this man can't play football."'

As it transpired, Kanu did play again, albeit not until April 1997 after having an aortic valve replaced in November 1996. And he kept on playing until his retirement in 2012, although he underwent further corrective heart surgery in 2014.

Injury and illness are part and parcel of a professional footballer's career, though try telling that to Kanu when his condition was first diagnosed or, more recently, to Watford's Younès Kaboul (263 days), Brighton's Steve Sidwell (250 days) or Bournemouth's Tyrone Mings (226 days), who finished first, second and third in the Premier League's 'days out injured' chart for the 2017–2018 season. Still, those stuck on the sidelines should be comforted by the thought that, thanks to advances

Top of the crocks

The table opposite ranks the 20 Premier League teams by the number of days their players lost to injury during the 2017–2018 season.

Watford's injured players spent the most days on the sidelines during the 2017–2018 season, which partly told the story of their late-season freefall. In mid-October, the Hornets sat in fourth place after winning four, drawing three and losing one of their first eight matches. Their 15 points equated to an average of 1.875 points per game. By the time of their season-ending loss – 1–0 away to Manchester United – they'd dropped to 14th on 41 points, and their points-per-game average had plummeted to 1.08.

Many blamed the board, because of their decision to sack manager Marco Silva in January. But was the decline more to do with the number of injuries? Then again, was it a new manager and different training methods that led to the injuries? No one will ever know for sure, but that won't stop fans arguing about it.

Staying fit and healthy didn't have the desired effect for Southampton, whose players spent the least time in the physio room but only avoided relegation on the final day of the season. If injuries weren't to blame, what was? Managerial instability – Claude Puel sacked in June 2017, Mauricio Pellegrino sacked in March 2018 – didn't help, neither did an apparent inability to pass the ball forwards.

▶ Watford's 2017/2018 season derailed for many reasons including the worst injury record in the league

in sports medicine and collective best practice, more players than ever before are returning strong from the most damaging of injuries.

But it's not just cutting-edge innovations like anti-gravity treadmills that are prolonging players' careers. What footballers put into their bodies is arguably of greater importance. Ryan Giggs gave up alcohol and fast food in his early twenties and ended up playing over 1,000 times for Manchester United. In 2017 the teetotal Jermain Defoe credited his switch to a vegan diet as one of the reasons behind his return to the England squad at the age of 34. And, as we'll see in the next chapter, clubs are spending millions on nutrition to increase players' durability and help them maintain a healthy weight and run faster.

INJURY RANKING (WORST TO BEST)	TEAM	INJURIES	DAYS MISSED	AVERAGE DAYS PER INJURY	FINAL LEAGUE POSITION
1	Watford	49	1,987	78	14th
2	Everton	56	1,822	73	8th
3	Crystal Palace	57	1,821	71	11th
4	West Ham United	73	1,627	60	13th
5	Manchester United	60	1,431	57	2nd
6	Burnley	40	1,415	57	7th
7	Liverpool	78	1,368	56	4th
8	Huddersfield Town	47	1,338	56	16th
9	West Bromwich Albion	57	1,208	56	20th
10	Swansea City	42	1,183	52	18th
11	Bournemouth	56	1,166	50	12th
12	Stoke City	52	1,152	49	19th
13	Newcastle United	45	1,092	49	10th
14	Tottenham Hotspur	49	1,038	47	3rd
15	Manchester City	50	970	45	1st
16	Chelsea	56	941	42	5th
17	Arsenal	71	909	40	6th
18	Brighton and Hove Albion	26	882	40	15th
19	Leicester City	40	836	38	9th
20	Southampton	38	618	26	17th

6
14

FUELLING PEAK PERFORMANCE

'Some players will have a glass or two of wine if they're on an international break [and not playing], but as far as their weekly fuelling goes, it just doesn't happen – it'd soon get sussed out due to the demands of what is required on the pitch.' The words of Celtic's head of sports science, Jack Nayler. Here's former England manager Roy Hodgson: 'I remember we looked to celebrate qualifying for the European Championships as we'd won 10 out of 10, so we got the champagne out. About four players took a glass while the rest wanted a can of coke.'

◀ Players like Southampton's Oriol Romeu are now fuelled by the best chefs and nutritionists in the world

It's a far cry from the days when 1970s terrace heroes like Rodney Marsh and George Best would meet in the pubs and clubs of London's King's Road. The man widely credited for footballers going dry is Arsène Wenger, who put a stop to Arsenal's drinking culture and what the players called the 'Tuesday Club'. 'On a Monday, Wednesday and Thursday people were dressed in tracksuits,' former Arsenal midfielder Ray Parlour recalled on online sports chat show *Sportlobster TV*. 'On a Tuesday everyone turned up with their suits on. George [Graham, then manager] was trying to work out what was going on – but we were obviously going out on the town.' That imperfect fuelling plan stretched to food, so much so that former Newcastle United striker Micky Quinn called his autobiography *Who Ate All the Pies?*

> The players here are as diligent – if not more so – as some of the best Olympians.
>
> **MIKE NAYLOR** CONSULTANT PERFORMANCE NUTRITIONIST, SOUTHAMPTON

'I'm asked about footballers and how committed they are to their diets all the time,' says Mike Naylor, the English Institute of Sport's head of performance nutrition, who also works for Southampton. 'I work across many sports and the players here are as diligent – if not more so – as some of the best Olympians. They see how important it is.'

Liverpool's finest signing

▲ Liverpool's head of nutrition, Mona Nemmer, has transformed what the club feeds its players

Liverpool's 2016 pre-season tour of the United States saw the Reds serve up a 2–0 victory over AC Milan, sandwiched between defeats to Chelsea and Roma. No long-term impact, instantly forgettable. But more significant developments were occurring off the football pitch. As the players prepared for their 10-month 2016–2017 campaign under the Californian sun in Palo Alto, manager Jürgen Klopp introduced one of the club's most important signings of recent times. Not paid a weekly six-figure salary, not attracting a million-plus Twitter followers, meet Mona Nemmer, Liverpool's head of nutrition, the woman who previously transformed the fuelling plans of Bayern Munich under Pep Guardiola and Germany's under-21 team.

'I'm not a classic nutritionist,' Nemmer says at Liverpool's Melwood training ground in an accent that's acquired a hint of Scouse. 'My aim is to match the theoretical side – the science – and the practical side. While it's true that players don't need to know the science behind the philosophy, they are increasingly interested.' Nemmer's academic background is nutritional science, but she went on to serve a chef's apprenticeship. That gave her the kitchen skills to put her nutritional know-how into practice. 'And I'm still studying, attending a course in Austria,' she continues. 'Nutrition's a fast-moving field and you need to keep ahead.'

Nemmer works with a team of chefs to provide individualised meals for the players all week round. Where many lower-league clubs have the resources for lunch at a push, Liverpool provide breakfast, lunch, an afternoon snack and, when needed, dinner. 'It gives us a chance to serve them as holistically as possible and fuel their individual needs.'

Clearly, Big Brother is at work again here, with clubs anxious to keep their valuable assets on the straight and narrow. But for experts like Nemmer and her fellow nutritionists working across football, the motivation is less commercial. Their aims include boosting immunity, aiding recovery from injury and maintaining peak body composition. Core to Nemmer's holistic outlook is the Liverpool canteen – or marketplace, as she puts it. Deep-frying is banned and, where possible, everything is fresh, seasonal and local.

'Product quality is so important in our system that we even have our own 1,200m^2 allotment near Anfield. We grow loads of different vegetables there including broccoli, carrots … we even have hives to produce honey. There are polytunnels and a little house in the middle where you can chill or sit down and have team events. It's really nice to say the Caesar salad we serve is made from Romaine lettuce from the allotment.'

Liverpool's captain, Jordan Henderson, praised Nemmer's impact in an interview with the *Liverpool Echo*: 'We all ate well anyway – as professionals you need to make sure your diet is good. But since Mona's come in, it's gone to another level – the food we've been having is unbelievable ... we're getting a bit spoilt, actually. I've never seen anything like it since I've been involved in football.'

Periodisation of carbohydrates

In years gone by, professional footballers regularly refuelled on fish, chips and mushy peas. Then, in the 1970s, many sports teams switched to a carbohydrate-heavy diet based on International Olympic Committee (IOC) guidance. Now Liverpool and other top-level clubs follow a carbohydrate-periodisation model. The theory derives from a significant body of research, including studies by former Liverpool nutritionist James Morton, who switched to cycling's Team Sky in 2015.

'The purpose of this nutritional strategy is to maintain body composition as well as promoting certain physiological adaptations [more of which later] via increasing and decreasing carbohydrate intake depending on the demands,' explains Morton. 'Here's an example for a team playing one game in a week. On Saturday you'd play the game, so training load is high, meaning you crank up the carbs. On Sunday you'd do the same to recover from the game. Monday might be a really light training session where you'll cover no more than 2.5km, so you don't need a lot of carbs for that. Tuesday or Wednesday might be the hardest sessions of the week, but still cover only 5–6km so they don't warrant a huge intake of carbs. Thursday might be more of a technical session so, again, low on carbs. You'd then go high again on Friday to prepare for Saturday's match. Essentially, you have low-to-moderate carb intake for most of the week but crank things up from match day -1 to match day +1.'

Morton implemented a traffic-light system – which he's carried over to Team Sky – where green denotes a high-carb day, amber a moderate-carb day and red a low-carb day. When teams are regularly playing three games a week, like Liverpool did in 2017–2018 when they reached the Champions League Final, the need to pack the cells with energy-providing glycogen is vital, so conceivably every day will be a green day.

Theory is one thing. Making it work is another matter. Morton concedes that his carbohydrate-manipulation model was difficult to implement because the players would be in for breakfast and lunch only. Nemmer and her team have sought to overcome this problem by providing all the meals of the day, starting with breakfast served at Melwood from 7.30 a.m. to 10.30 a.m. every day. This early start ensures that players have a minimum gap of 90 minutes between eating and training, though they might do some low-intensity activation work with the physios beforehand.

> We also use many herbs and spices, because the bigger the flavour, the better for your metabolism.

MONA NEMMER HEAD OF NUTRITION, LIVERPOOL

A feast for the eyes

Nemmer is keen to add a touch of theatre to proceedings – this is to draw players in rather than to seek applause. 'At breakfast we might have an omelette station or a porridge station where players can choose rice milk instead of normal milk. They can also add roasted apples as a topping. Or half a banana, served by a lovely person saying, "What can I do for you this morning?"'

Muesli and nuts are also an option and, like a growing number of clubs, including Everton and Southampton, Liverpool have a smoothie station kitted out with NutriBullets® and juicing machines. Here, chefs – or the players themselves – will blend a mix of fruits, vegetables and seeds.

'We have recipes based on key words, like "recovery" and "immunity boost", which gives the players guidance and education without pushing it too hard,' Nemmer says. 'But we like to mix it up. So they might have carrot, apple, orange or blueberry and strawberry, which they enjoy. Or a mixture of celery, peach and apple. It's their choice. We also use many herbs and spices, because the bigger the flavour, the better for your metabolism.'

Recovery smoothies often include pineapple juice, because pineapple contains bromelain, an enzyme that helps to break down and digest proteins to repair muscle.

Another popular ingredient is cherry juice. 'If players have high muscle soreness, which they might after a game, we'll include a relevant smoothie station,' says Lloyd Parker, Everton's performance nutritionist. 'It'll feature lots of berries but also added CherryActive®, which is a concentrated form of cherries. Certain foods, like cherries, reduce muscle soreness due to their anti-inflammatory properties.'

Smoothies provide a carbohydrate hit and, with some nutritional alchemy, Nemmer can also control energy release. 'If you mix fennel in your apple, orange and mango juice mix, the whole

Beetroot boost

The performance-enhancing properties of beetroot juice have received a great deal of press these past few years. When beetroot is digested nitrates within the vegetable are converted to nitric oxide. Studies have shown that this process reduces the oxygen cost of low-intensity exercise and extends time to exhaustion in high-intensity exercise (in other words, you can keep going longer). These benefits have been found to be greater for the recreational athlete than for the elite. Researchers suspect that this is because, to put it bluntly, amateur sportspeople have more room for improvement. So if your level is more Sunday league than Champions League, a beetroot shot before a match could stop you hitting the wall after 70 minutes.

▲ It seems Ajax striker Kasper Dolberg isn't powered by smoothies

thing is broken down more slowly, so will provide a slower delivery of energy [which helps the players fuel for the upcoming training session]. All of this information is displayed on posters on the walls so the players are educated and aware.'

The posters also explain that smoothies and juices are a great way to deliver daily antioxidants (see box overleaf), because they provide a big nutrient hit in an easily digestible form. This is a relatively accepted model in nutrition, but there are sceptics who question whether a smoothie is the optimum mode of delivery. Over in Amsterdam, while Ajax nutritionist Peter Res offers star striker Kasper Dolberg and co. the smoothie option, he does so with a sprinkling of caveats.

> Certain foods, like cherries, reduce muscle soreness due to their anti-inflammatory properties.
>
> **LLOYD PARKER** PERFORMANCE NUTRITIONIST, EVERTON

'I'm not a big fan of smoothies,' he says. 'I believe that we as humans evolved to chew on food, for saliva to break things down. A smoothie is almost like a fast food. Our brains can't register it so people tend to overeat or, in this case, overdrink because it goes through too quickly.'

To understand Res's point we need to consider how our bodies 'know' that we have eaten enough. This comes down to the interplay between the

Individual dietary needs

FOOTBALL CLUBS HAVE TO CATER FOR AN ARRAY OF INDIVIDUAL DIETARY REQUIREMENTS

Arsenal right-back Héctor Bellerín made headlines at the start of 2018 when he revealed he'd been following a plant-based diet since the start of the season, claiming that cutting meat and dairy had improved his performance. He also said that his new diet helped to prevent injury. 'For me the most important thing is the [reduced] inflammation in my body after games and the speed that my body recovers compared to before,' Bellerín said at the time.

Bellerín certainly started the 2017–2018 season in peak form, albeit there's no concrete evidence that veganism or vegetarianism can cut injury risk. In fact, there's an argument that dropping meat could leave footballers missing out on certain nutrients; good-quality meat is a particularly rich source of proteins, vital in muscle repair.

Vegetarianism isn't the only dietary requirement clubs need to cater for. 'We have some players who follow a gluten-free diet so our chef will also cook gluten-free pasta, for example,' says Peter Res, nutritionist at Ajax Amsterdam. In January 2018, while still at Arsenal, Jack Wilshere revealed that he'd moved to a gluten- and dairy-free diet to help him lose weight.

'It gives me belief in my body,' Wilshere told the *Telegraph*. 'I know the right foods to eat; I know the best way to recover and get the right amount of sleep.'

The gluten-free diet first came to prominence as a means of treating coeliac disease, a digestive disorder in which the body is abnormally sensitive to gluten, a protein component of wheat, rye, barley and crossbreeds of those grains. But there has been an explosion in non-coeliac followers of the diet, who turn to it, as Wilshere did, to lose weight. The gluten-free diet is also associated with reduced inflammation throughout the body.

Despite the diet's popularity, there's very little empirical evidence that going gluten-free is beneficial to an athlete. In one study, researchers at the Canadian

◀ Many clubs, like Ajax, will cook halal for players like Hakim Ziyech who are Muslim

Sport Institute Pacific tracked two groups of non-coeliac competitive cyclists, one group following a gluten-free diet and the other retaining gluten in their diet. The findings showed no difference in athletic performance, perceived gastrointestinal distress, inflammatory markers, intestinal damage or overall well-being between the two groups.

One theory is that any benefits come not from the elimination of gluten itself, but from the fact that followers of the diet eat less refined carbohydrate, which is of limited nutritional value, and more nutrient-rich whole foods and fruit and vegetables.

Of course, not all dietary requirements are health-based. A club's chef must also be aware of players' religious and cultural backgrounds. 'We have a lot of Muslim players, so everything we make is halal,' adds Res. 'We don't have pork at the club – maybe the occasional slice of ham is available, but never in a main meal. It's easily replaced by fish, chicken or veal, so we don't need it anyway.'

hormones cholecystokinin, released by the intestines, and leptin, released by fat cells, which tells your brain about your long-term needs. Leptin amplifies the signals that cholecystokinin sends to enhance your sense of fullness and also works with the neurotransmitter dopamine to give you feelings of pleasure after eating. The problem is that this process takes around 20 minutes. Down a smoothie and you might then launch into a big breakfast, unaware that the smoothie has actually fulfilled all your nutritional needs – your body just hasn't realised it yet.

Res presents the science in a more digestible package to Ajax's youth players. 'We have workshops and I'll show them three oranges next to a glass of orange juice. I say, "OK, can someone eat these three oranges and then have dinner?" They look at me like I'm crazy. But I explain that this is what you drink next to a meal if you have an orange juice, and, ultimately, that's extra calories.'

Lunch is served

Once the Liverpool players have completed their training session, it's back to the canteen for lunch. Again, this is where Mona Nemmer's experience as a chef pays dividends. 'Lunch is like a marketplace and there's a lot of interaction between the players and chefs; it's vital you present food in an attractive way. So we'll offer a lovely piece of fish grilled in front of the player. Throw in a mixed salad from our allotment or pasta fresh out of the pan. There's more theatre to it, as well as tailoring everything to player preferences and nutritional needs.'

As former Liverpool nutritionist James Morton highlighted, how much of each macronutrient – carbohydrate, protein and fat – a player needs is dictated by the intensity of the training session or match. Nemmer and Res both explain that they break the plate down into these three main groups with the proportion of each shrinking or rising depending on where the player is in the week (and the intensity of the sessions). That's where signage helps; there are posters dotted around the canteen to educate the players. It's the same for Mike Naylor at Southampton, who has installed monitors that constantly relay messages of nutritional support like 'protein foods, like tuna and chicken, boost muscle repair'.

Clubs might have a double training day or certain players might stay in for extra gym work. At Liverpool, Nemmer and her team provide for these situations by offering a mid-afternoon snack of a 'little rice dish or something like a yoghurt with nuts

> I've even visited Winchester sushi bars that are popular with the team to tell the chefs which are the best foods on the menu for the players!

MIKE NAYLOR PERFORMANCE NUTRITIONIST, SOUTHAMPTON

sprinkled on a homemade muesli bar'. Down at Southampton, the players 'go mad for beef jerky', says Naylor. 'But that's good as it's high in protein; we look at high-protein options in the afternoon and the chef does lots of snack pots, with something like a small portion of chicken breast and veg to take away. It helps with recovery.'

Unlike most clubs, Liverpool also lay on dinner as, says Nemmer, the players often train late into the afternoon. It's something other club nutritionists must look at with envy. 'We've calculated that the players have 2,100 feeds each year, including snacks, and only about 740 of those feeds are at the club,' says Southampton's Naylor. 'You can only control so much.' Don't feel for the poor, neglected players too much, though – the money swilling around the game means many have private chefs.

Manchester City players Kevin De Bruyne and Ilkay Gündoğan employ Jonny Marsh, a Michelin-starred chef. 'I make food at home and drop it at Kevin's house twice a week, so it's ready in his fridge,' Marsh told the BBC. 'With Ilkay, I'll go round to his house every night and cook.'

Then there's Newcastle United midfielder Jonjo Shelvey, who in July 2015, while playing for Swansea City, advertised for a personal chef. Shelvey was offering a salary of up to £65,000 per year to the successful applicant.

Manchester United's young striker Marcus Rashford is catered for by performance food outfit Nourish Fit Food – a bespoke meal delivery service based in Cheshire. There can't be too many 20-year-olds like Rashford who spend their Saturday nights nibbling on 'protein flapjacks'.

▼ Liverpool provide breakfast, lunch and also dinner for players training late in the afternoon

The pros of antioxidants

We often hear about antioxidants being a 'good thing', but what exactly are they? Put simply, antioxidants are substances that protect your cells against the harmful effects of free radicals. These are unstable molecules produced when breaking down food or exercising hard – in essence, a waste product from metabolic reactions. They are unstable because they contain atoms with unpaired electrons. Electrons like to be in pairs, so these free radicals scavenge the body to find partners for their unpaired electrons. This sets in motion a chain reaction, which can damage cells, proteins and DNA; high levels of free radicals are blamed for conditions ranging from cancer to Parkinson's disease. Antioxidants keep free radicals in check by donating an electron without becoming destabilised themselves, and so prevent the free-radical chain reaction.

'Of course, often the player or their partner might do the cooking themselves,' adds Nemmer, while eating out is common – with guidance, of course. 'We give the Southampton players a list of the most appropriate places to eat,' says Naylor. 'I've even visited Winchester sushi bars that are popular with the team to tell the chefs which are the best foods on the menu for the players!'

Pile on the protein

Club nutritionists are called upon to fill the players' minds as well as their stomachs. They are available to clarify and correct claims made on social media and by the wider press about some new game-changing food or diet. Spending more time inside the training centre rather than out on the pitches, players recovering from injury are among the most available and receptive students. 'Injury is obviously not great for the player or the team, but it gives you time to have a two-way nutritional conversation,' Naylor says.

'Beyond that, the nutritional strategy changes for an injured player,' he adds, alluding to two key requirements: increased protein intake and foods that support inflammation management. 'It means I'll be nudging the chef to include certain foods high in protein, like lean chicken and fish, and environmentally nudging the player by setting up a protein station earlier in their buffet selection. That kind of approach doesn't just apply to injured players, of course. If it's a high-carb day, we'll position carbs first; if it's a lighter-carb day, we'll start with the salads, veg and meat and then have carbs at the end of the line.'

Protein, in particular, plays a key nutritional role whether a player is injured or not. Proteins comprise a combination of amino acids. Some amino acids, such as alanine, serine and glutamic acid, can be made by the body; others, known as essential amino acids (e.g. leucine, lysine and tryptophan), cannot. Therefore, it's vital adequate protein is ingested daily to support protein synthesis (the process by which individual cells build their specific proteins) and promote recovery.

Footballers generally need around 1.4g protein per kilogram of bodyweight. For an 80kg player like Cristiano Ronaldo, that equates to 112g protein, but timing is often more vital than quantity. Research shows that resistance-training adaptations, like increased muscle strength, are more effective if a small amount of protein (6–10g) is consumed 15 minutes before the session. Around 200g yoghurt or 300ml

APPLY THE SCIENCE **Understanding vitamins**

Vitamins are essential for maximising training gains. Here are some of the key ones

Vitamin C Vitamin C is best known as the first defence against colds and upper respiratory tract infections – common when overexerting yourself – but research is equivocal on its effectiveness in this respect and it is particularly unclear whether heavy doses kill colds any faster than the NHS's recommended daily amount (RDA) of 40mg. However, it is well established that vitamin C can help strengthen capillary walls and blood vessels for better bloodflow – useful for footballers, who cover upwards of 10km each match. 'It also improves iron absorption, which has clear footballing benefits as iron helps oxygen bind to blood that's then delivered to working muscles,' says nutritionist Lucy-Ann Prideaux. There's an argument that you shouldn't take vitamin C straight after training as research shows it can blunt the adaptation process. Just half a red pepper or one large orange will provide the recommended amount.

Vitamin B1 (thiamin) Vitamin B1 could be the footballer's most vital vitamin as it plays a pivotal role in converting glucose into energy (as do vitamins B6 and B12). It also strengthens the nervous system for more efficient sprinting. The NHS recommend a daily dose of 1mg of B1 for a man and 0.8mg for a woman. Vitamin B1 is found in wholegrains, nuts, pulses, pork, fruit and vegetables, but as vitamin B1 is prone to heat destruction through cooking, a supplement might be needed.

Vitamin E The antioxidant vitamin E helps to sweep up free radicals (see page 107), so it's key to strengthening the immune system. Like vitamin C, though, there's evidence that taking in too much vitamin E can hamper cellular adaptations in the working muscles. The RDA is 4mg, which is easily ticked off with around 150g of almonds or a healthy serving of spinach. As vitamin E is stored in body fat, and not lost in urine, a supplement is not usually necessary.

milk equates to 10g protein, although players often consume a protein-based energy gel instead.

Most players are also encouraged to consume the textbook optimum of 20g protein after the session to stimulate muscle-fibre repair, though that theory too is open to debate and interpretation. 'That 20g figure comes from a [2009] study by Stu Phillips [Professor of Kinesiology, McMaster University, Hamilton, Canada] where they used a small muscle group and discovered 20g maximised protein synthesis and muscle repair,' explains physiologist and nutritionist Asker Jeukendrup, who has worked with Chelsea and currently guides RB Salzburg, PSV Eindhoven and FC Barcelona. 'I suggest it should be higher – maybe 30g – as football engages large muscles like the hamstrings, glutes and quadriceps.' Phillips himself has since revised his protein guidance, suggesting an intake of 0.3–0.4g of protein per kilogram of bodyweight. 'We appreciate that body size has to be part of the equation,' says Phillips. The actual quantity may be open to debate, but the recovery benefits of protein aren't, which is why Jeukendrup, Phillips and their counterparts recommend a protein shake straight after a match (see chapter 6).

More good fats, fewer bad carbs

Let's not forget fats – or good fats. Extra virgin olive oil is a staple at every football canteen table. That's because fats play a vital role in forming new cells. And that's from the cells that make up organs to the organelles within cells, which includes the powerhouses known as mitochondria. Better fats create better conditions for mitochondria to provide the footballer's muscles with the energy they need through a stop-start game of football.

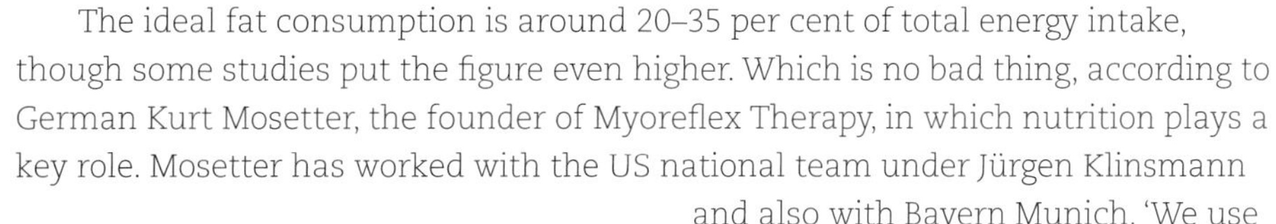

The ideal fat consumption is around 20–35 per cent of total energy intake, though some studies put the figure even higher. Which is no bad thing, according to German Kurt Mosetter, the founder of Myoreflex Therapy, in which nutrition plays a key role. Mosetter has worked with the US national team under Jürgen Klinsmann and also with Bayern Munich. 'We use a well-directed "glycoplan" as it has the potential to transform a player's health and performance,' explains Mosetter. 'It's currently being used by TSG Hoffenheim to great effect.'

▼ Bayern's Arjen Robben tucks into a meal packed with omega-3s

The glycoplan removes poor carbohydrates (foods high in sugar and flour) from the diet and instead prioritises high-quality, healthy fats. It's all designed to improve the body's anti-inflammatory system and, in Mosetter's words, 'economically educate the metabolism'.

'I call it "natural eating" and it's something I've learned from my travels in Nepal and studying biochemistry. Modern humans have been around for hundreds of thousands of years and in the majority of that time wholegrains provided less than 1 per cent of their dietary needs. Then from 10,000BC to now that figure grew to something like 10 per cent. That's not great from a metabolic viewpoint and is why I focus more on good fats like those found in oily fish, meat, coconut water and oil, nuts and almonds.'

Carbohydrates are part of the diet, but only so-called 'good carbs' like buckwheat and quinoa. Bread, noodles and pasta are avoided because 'they stimulate insulin and higher blood sugar levels. That's fine for a short burst of exercise but we want our players to perform to their optimum to the final whistle.' This principle will be familiar to anyone who has ever followed a GI diet, which seeks to minimise consumption of foods with a high GI (glycaemic index), which cause spikes in blood sugar levels, in favour of low-GI foods, which have a more gradual effect.

Mosetter's glycoplan has parallels with the ketogenic diet, which again is high in fat and low in carbohydrate so that the body is forced to burn fat for fuel rather than carbs. The ketogenic diet has been more popular in endurance sports like distance running and cycling than in football, which, as we have seen, has a high anaerobic element to it, requiring sudden bursts of action. However, Mosetter sees things differently. 'It's all about re-educating the system,' he says. 'There have been studies showing that athletes on ketogenic diets actually store greater levels of glycogen [the muscle fuel normally derived from carbs].'

That's nectar when it comes to repeated sprinting – and it's just as well, as the

▼ Juventus' Sami Khedira followed the 'glycoplan' after the 2016 European Championships

> Modern humans have been around for hundreds of thousands of years and in the majority of that time wholegrains provided less than 1 per cent of their dietary needs.
>
> **KURT MOSETTER** FOUNDER OF MYOREFLEX THERAPY

traditional natural high-sugar source is also off the menu. 'Bananas disturb the gut and cause a rise in uric acid,' Mosetter says. 'This causes a rise in ammonia. And if ammonia's rising, cognitive performance decreases.'

Like Hoffenheim, RB Leipzig are disciples, as is the German midfielder Sami Khedira. 'He followed a strict version of the glycoplan after the 2016 European Championships in the off-season, so no carbohydrates for one week,' Mosetter says. It would appear to be doing him no harm, as Khedira is thriving at Juventus and even scored his first ever hat-trick at the age of 30 in October 2017!

Adapting endurance

Mosetter's strategy may seem extreme and there are critics who insist high-intensity activity like football demands a high intake of carbohydrates, particularly before and after a match. However, as we have seen (see page 101), there is a growing body of evidence that suggests carbohydrate restriction at certain times during the training week can be of benefit, so it would appear that the world of football nutrition is moving closer to Mosetter's basic tenet of high-quality fats and low carbs.

In 2013 former Liverpool nutritionist James Morton led a study in which a group of young footballers were made to fast in the evening and morning before a high-intensity session. Soon after, the process was repeated with the same group, but this time they ate carbs the evening before and, the next morning, immediately before, during and after the high-intensity session. The results were startling and made many question perceived wisdom.

'We took muscle biopsies of each footballer and measured the glycogen stores and, understandably, they were lower in the low-carb group,' Morton explains. 'However, things were very different when we measured the protein activity...'

Morton discovered that the response of two key proteins involved in favourable endurance adaptations was enhanced rather than inhibited by restricting carb intake, whereas textbook carbohydrate intake actually blunted the protein response. In short, dropping the carbs had a stronger endurance effect than training with carbs.

This phenomenon is known as mitochondrial biogenesis, and as a result of these changes, the footballer becomes more efficient at using fat for fuel at a given exercise intensity, which means they produce less lactate – and fewer fatiguing hydrogen ions – so they conserve glycogen. Morton doesn't advise depleting glycogen levels during the season, especially when a team's playing twice a week, but it certainly has potential in the off-season, a time when clubs are not only looking to raise the aerobic capacity of players but also ensure optimum body composition.

Lean, fit and ready for action

It's common for gyms at football clubs to feature whiteboards displaying a number of fitness-related scores for each member of the squad. These might include key footballing attributes like jump strength and sprint speed. You will also often find body-composition results, taken regularly by the traditional method of skinfold – 'pinching an inch' with a set of callipers. However, it can take up to an hour to conduct a fully accurate skinfold test, so clubs often do a quicker version more as a snapshot than as 100 per cent truth.

Accuracy is not an issue at the Manchester Institute of Health and Performance. 'This is our DXA machine,' performance lead Rachel Howe explains, as she shows me around the technically advanced centre on the Etihad Campus. 'It measures body composition in specific areas, using a traffic-light system to display results: red is for a high-fat area and green is for a low-fat area. Each player takes around seven minutes to scan and you'll often have teams in here every six weeks or so. You can track whether fat's been lost in a certain area, for example. It can also pick up muscle repair, so it's used to monitor recovery from injury.'

DXA (Dual-energy X-ray Absorptiometry, pronounced 'dexa') is regarded as the industry gold standard for measuring body composition; the International Olympic Committee state that it's their 'method of choice'. A DXA machine is similar to an MRI scanner in that you lie down while you are scanned from head to toe. It passes two low-dose X-rays through your body. Different body components – bone, muscle mass, fat – block different amounts of radiation, so by detecting how much radiation passes through each part of the body, the scanner builds up a detailed picture of your body composition.

▶ Giant among giants: Harry Kane and his lean, powerful team are a sign of the times

'It gives you so much useful information,' explains Dr Karen Hind, bone and body composition lead at Leeds Beckett University. 'The data can inform mid-season monitoring, pre-season conditioning and injury rehabilitation.'

Hind says that it's so detailed it can even identify muscle asymmetry between limbs, which can help to predict injury. She is currently undertaking a project to determine 'body-composition reference ranges for English professional footballers'. 'So far players have come to this or other universities but I've just co-invested in a mobile DXA scanner so football clubs don't need to travel to us anymore. That was one of the deterrents for many pro clubs.'

Despite some clubs' reluctance, Hind has racked up 189 scans in what she believes is the first DXA study of its type anywhere in the world. The study is ongoing so those numbers will continue to grow, but currently the standout line is that goalkeepers tend to have the highest fat mass with very little difference between the other positions.

More specifically, the 18 keepers weighed an average 85.6kg with proportion of body fat at 17.5 per cent. That compared to the 58 defenders (68.1kg, 14.1%); 57 midfielders (58.0kg, 14.2%) and 56 attackers (67.9kg, 13.9%).

A different game

Barry Drust is one of British football's first sports scientists and has worked with England, Middlesbrough and Liverpool. He's currently professor of applied exercise physiology at Liverpool John Moores University and over the years has observed the changing shape of the Premier League.

'When I started, central defenders measured 6ft 6in and weighed 100kg; the centre-forward might look similar. But wingers and central midfielders would look very different – smaller and nippier,' he says. 'Now look at a team like Tottenham. Scan them all and can you work out who plays where? It's not possible. Anthropometrically, visually, they just look the same: stronger and leaner, more like athletes.'

> Look at a team like Tottenham. Scan them all and can you work out who plays where?
>
> **BARRY DRUST** SPORTS SCIENTIST

This reflects the growing physicality of the game in this country, which we focus on in chapter 1. The financial imperative of remaining in the Premier League means clubs are paying greater focus to areas like nutrition.

In an effort to stay at the top of their game, individual players might also seek out experts beyond the club. In

2015, five-time World Footballer of the Year Lionel Messi made headlines in Italian newspaper *Corriere della Sera* for his work with Dr Giuliano Poser, a sports nutritionist who specialises in a complementary therapy known as applied kinesiology. Its principles revolve around optimising muscular performance through the use of specific foods. Messi sought Poser's help to conquer vomiting urges before matches and various niggling muscle issues.

▲ According to reports, Dr Giuliano Poser helped Lionel Messi conquer his vomiting urges

Using a 'complex mathematical formula', the Italian physician measures how muscles react to specific foods. So out went sugar, refined flours and in came organic, seasonal food. 'I see myself as simply making fine adjustments to a powerful supercar by encouraging the use of organic food, raw grains, seasonal fruits and vegetables, extra virgin olive oil, eggs and fresh fish.' Poser is also credited with helping Juventus striker Gonzalo Higuaín cut the pounds.

Ultimately, whatever complex or simple method a nutritionist employs, key to maintaining a 'footballing body composition' is balancing calories in and calories out. Liverpool John Moores University PhD student Liam Anderson spent every day at Anfield for three years to measure and analyse the energy intake and expenditure of six unidentified Liverpool footballers.

'It was the first time someone had done this with a Premier League team,' says former Liverpool nutritionist James Morton. 'It was no mean feat – Liam even had players taking photos of their meals and sending them to him to analyse. The players loved him and wanted him to succeed.'

Anderson used what's deemed the most accurate method available to measure energy expenditure, namely the 'doubly labelled water' method. This technique uses varying excretion rates of oxygen and hydrogen to calculate calories burnt.

To measure energy intake, he collected detailed nutritional information from the six players on different days: match days, recovery days and training days.

'The results were positive with the only real recommended change around the distribution of carbohydrate intake,' says Morton. 'It was higher at dinner; ideally you'd take in more carbs earlier in the day so you can use them for training.'

Anderson is now working as head fitness coach for Videoton in Hungary, whose manager, former Blackburn and Manchester United defender Henning Berg, had asked Morton for a recommendation.

As Drust pointed out, today's footballers are professional athletes whose clubs spend millions of pounds on educating and feeding them to perform at their optimum. When did you last hear the classic 'Who ate all the pies?' chant during a Premier League match? You suspect that the likes of Jan Mølby and Neil Ruddock, no matter how talented, would simply lack the speed, strength and endurance to keep up with elite football in 2019. While the wider press rejoices in slow news day morsels like Antonio Conte banning ketchup and Pep Guardiola banning chocolate, the truth is that a top footballer's everyday nutrition plan is much more scientific than that. It's a meticulous approach based on empirical evidence, which carries over to the match-day menu.

Recipes for success

COLOUR-FIT IS A SIMPLE BUT INGENIOUS WAY OF HELPING FOOTBALLERS MAKE BETTER FOOD CHOICES

'Necessity is the mother of invention' is the old English proverb that could have been written for Colour-Fit. No matter how much Preston North End's head of fitness, Dr Tom Little, worked with his players on nutrition, they still made poor choices.

'We'd do presentations, put up posters, take them shopping, give them cooking lessons – but whenever we analysed what they'd been eating, it was still pretty rubbish,' Little told football website Training Ground Guru, 'and after all the work we'd done, I'd end up trying to throttle them, like Homer with Bart.'

It's why, Little tells me at the inaugural Soccer Science conference held in June 2018 at Bristol City's Ashton Gate, he devised Colour-Fit charts that divide meals into three 'food goals'. Green is 'fuel to maximise exercise performance through carbohydrates'; red is for 'lean muscle' – proteins to help players stay lean and aid muscle growth and repair; and gold is for 'health', achieved by eating foods rich in vitamins, minerals, fibre and good fats.

Together with his team, which now includes a full-time chef, Little has created a library of over 200 meals, each accompanied with a recipe card and short instructional video. There are also over 30 diet plans for different sports and physical goals.

The Colour-Fit app was officially launched in 2017. It's certainly proving popular with Premier League clubs, with eight clubs currently signed up for a whole-squad subscription, including Manchester United, Newcastle United and Arsenal. However, it's also available to individual subscribers, so if your nutritional approach is more KFC than LFC it might be something to take a look at.

sky
CHAMPIONSHIP
EFL
PP

MATCH-DAY FUELLING AND SUPPLEMENTS

Team buses very rarely hit the headlines. Why would they? They're buses. That briefly changed on an overcast Tuesday back in May 2016 when a lively West Ham crowd welcomed Manchester United's £400,000 45-seater Van Hool TDX27 Astromega with a brewery's worth of empty bottles and crushed cans. While Phil Jones and Adnan Januzaj took shelter in the aisle, Jesse Lingard spotted an opportunity to please his followers and shared his selfie video online. An act of violence diluted by likes.

Thankfully, part of that near half-million pound investment went on shatterproof double-glazed windows so the players remained unharmed. Sadly for them, United couldn't even enjoy the last laugh, losing 3–2 at the Hammers' farewell game at Upton Park before the move to the Olympic Stadium. Still, there was a crumb of comfort – within their 'battle bus', they could enjoy quality meals on wheels thanks to a kitchen equipped with two Neff hot-air ovens, a Panasonic microwave, Kenco single self-serve coffee machine, refrigerator, as well as hot-water boiler tap, a coffee percolator and catering facilities. Where once elite players hopped on a coach loaned from National Express, now an increasing number of clubs – Bayern Munich, Barcelona, Real Madrid – are delivering their assets in bespoke style.

'We took delivery of a new team bus at the start of the year and I'm very proud of it,' says Liverpool's head of nutrition, Mona Nemmer. 'I helped to design it and it features everything you need, especially fridges, which are particularly important because we cook food before we leave and take it with us. We then reheat the food on the team bus for after the match. There's a little board that we write the menu on – stuff like pasta or quinoa. We prep the salad on the bus during the match.'

◀ Gels, consumed here by Preston's Tom Clarke, are a staple energy source on match day

Bayern Munich's bus is equally versatile and was built by Munich firm MAN. It features card tables, which rise from the floor to adjust to the height of the player sitting there (although presumably there has to be some harmonious middle ground when Franck Ribéry, 1.70m, plays snap with Manuel Neuer, 1.93m), as well as leather seats, satellite TV and numerous 19-inch LED screens. There's also a fully equipped kitchen. So proud are the Munich club of their team bus that they unveiled it in 2014 to 300 VIP guests, who watched in bewilderment as it emerged from a garage, enveloped in clouds of dry ice.

Eating on the road

Team buses and their capacity to feed the players come into their own after matches. We'll delve deeper into the protocols and science later. But let's not get ahead of ourselves. First, players need enough fuel to see them through 90 minutes of play and two blocks of injury time – an average of 1,106 calories according to research undertaken with Serie A players. To ensure the players are fully fuelled takes significant planning and resources, especially when you consider a first team might transport 16 to 18 players to a match.

'We're lucky enough to have separate teams of home and away chefs, and that includes two away teams – one for European matches and one for domestic away matches,' Nemmer explains. 'You need this, really, to achieve some form of work–life balance. Take a midweek Champions League game. The chefs fly out on Sunday, preparing to feed the players who arrive on Monday. They'll feed them again on Tuesday, the match day, but, with the evening kick-off, inevitably the

◄ Away days see head of nutrition Mona Nemmer feed Liverpool players on the team bus

▲ Arsenal's Emirates Airbus A380 features a mini bar – alcohol-free, of course!

chefs don't fly back until the Wednesday. If you're then playing away on the Saturday you'd be driving off again on Friday and that's too much, so the domestic away team take over.'

Feeding players on long flights is one of the many logistical hurdles for the support team. 'We have influence over the plane's catering, but we can't physically prepare the food,' says Nemmer. 'But we don't need to as they know what standards we have.'

With all the money swilling around the upper echelons of professional football, the world's best tend to charter private jets decorated in the club's livery rather than have to use scheduled flights. Since 2011, Manchester City have flown in a specially decorated A330-200, while Arsenal travel to European away matches aboard an Emirates Airbus A380 plastered with images of Mesut Özil and co. Within the aircraft, each player has a bed, an alcohol-free mini-bar and personalised media console, and there are even on-board showers.

Most clubs playing abroad rely on the designated hotel to source the players' food even if, like Liverpool, they bring their own chefs. Again, that doesn't come without issues, particularly when it comes to the origins of meat.

'Legislation for use of hormones in cattle-rearing varies by country,' says nutritionist Peter Hespel, who works with the Belgian national team. 'That's how a hormone like clenbuterol, which is banned by WADA [World Anti-Doping Agency], gets into the food chain. It's not a problem for meat produced in the EU where any hormonal treatment's forbidden. But if the meat has come from, say, a South American country, there could be a problem.'

Clenbuterol is used in livestock to increase lean mass and raise the overall yield of production. Its muscle-building properties explain why it is classified as a performance-enhancing drug. So endemic is clenbuterol in South America's food chain that at the 2011 Under-17 World Cup held in Mexico, where Liverpool's Mona Nemmer was working with the German team, over 100 players tested positive for the banned drug. That same year, five senior players of Mexico's national team also tested positive for clenbuterol before the Concacaf Cup. (Mexican cattle ranchers are actually banned from using clenbuterol as a growth enhancer, but reports suggest it's still used widely.)

FIFA ordered meat samples to be collected from team hotels and 30 per cent of those showed the presence of clenbuterol. 'It's why we're particularly diligent in some countries we play in pre-season,' says Nemmer, possibly alluding to China, where clenbuterol is also used. Liverpool visited the Far East during the 2017 pre-season. 'I'll contact the hotels where we're staying six to eight weeks beforehand to pin down food sourcing,' she continues. 'One oversight could damage both my career and the player's.'

Lower-league clubs may not need to worry about South American clenbuterol, but their less exotic away trips present different challenges, such as how to apply nutritional best practice when you lack the resources of the world's top teams.

Bristol Rovers play in England's League One and are on something of a high after gaining two promotions in the past four seasons. Although their owner, Wael al-Qadi, from the family that founded the Arab Jordan Investment Bank, is wealthy, his favoured hashtag of 'evolution not revolution' makes it clear that he's clutching tightly to those purse strings.

'When we're playing away, we tend to eat pizza on the bus before the match,' says Rovers' head of medical services, Keith Graham. 'The players don't particularly enjoy it – they find it boring – but if you can tell me what company can provide

> I'll contact the hotels where we're staying six to eight weeks beforehand to pin down food sourcing. One oversight could damage both my career and the player's.

MONA NEMMER HEAD OF NUTRITION, LIVERPOOL

28 hot meals, which their employees can eat with their hands, at exactly the right time on a Saturday, unless you have your own staff ... well, you can't.'

Pre-match feeding

Hopefully, the Bristol Rovers players' pre-match pizza is a margherita made with white flour, as it's common practice to refrain from eating too much fibre shortly before a match to prevent gastric – and very public – distress. That means wholemeal rice will be replaced with white rice, brown pasta with white pasta ... but as nutritionists James Morton and Asker Jeukendrup emphasised in the previous chapter, high-carbohydrate feeding is key to prepare for the high intensity of a match.

'Our main pre-match meal works back around three to four hours before kick-off,' says Jeukendrup, an expert on carbohydrate feeding who works for RB Salzburg. 'That's where you'll top up your glycogen stores in the liver. I say liver because the glycogen stores in your muscle should be high from the previous evening's meal. Physiologically, this is because liver usually grabs the glucose first to store as glycogen. Anything left over goes to the muscle, and that stays there unless it's used.'

▼ Asker Jeukendrup introduced a science-led fuelling plan to RB Leipzig

So, unless the Leipzig players are vigorously sleepwalking around Saxony the night before a match, their muscle glycogen levels should be relatively stable and sufficient to sustain 11–13km of running? Not necessarily, according to Jeukendrup. 'When you sleep, the brain uses up a lot of glycogen, which is released from the liver. That leaves little in the liver. This must be replenished, but to achieve the full 80–90g capacity, you're looking to consume 200g carbs, because much of the carbohydrate will bypass the liver and be stored in other tissue.'

As decision-making is inhibited when brain glucose levels are low, you can see the importance of this pre-match feeding. 'It's not all about the liver, though,' adds Jeukendrup. 'At very high intensities, players will burn through muscle glycogen at a high rate – maybe 3–5g of carbs per minute.' That's where sports nutrition enters the match-day larder. Once lunch has been consumed and as the match approaches, players will often sip on an energy drink (more on fluid needs in chapter 8) or ingest an energy gel. These moderately viscous sachets of energy have proved popular over the past few years as they've actually become palatable. Years ago, squeezing out an energy gel was akin to mangling that last nodule of toothpaste. Now, manufacturers like SiS, former partners to Liverpool, produce more isotonic gels, which have a higher water content and are digested far more easily. Each gel contains around 20–30g fast-releasing carbs (see page 111) and will often be consumed during a warm-up on the pitch about half an hour before kick-off.

'They don't always need a gel before the start of the match,' says Liverpool's Mona Nemmer. 'If the tactics are to play in a pressing, high-intensity style from the outset, they might need one, yes. But often it's more for the head than anything else as they should be well fuelled by now.'

Nemmer's comments lend weight to Kurt Mosetter's 'glycoplan' concept, which we came across in the previous chapter. Mosetter argued that fast-acting sugars shouldn't be needed if the player's metabolic system has been gradually retrained through the introduction of more high-quality fats in the diet.

This issue of match-day feeding is an area Everton's Graeme Close investigated in a study of elite rugby league players. That's quite a coup for Close because elite sports people are rarely willing – and often unable because of time constraints – to offer themselves as guinea pigs. University sports science students are the traditional lab rats, and often professional clubs source studies from worldwide journals, then perform in-house tests to see how far the findings apply to their players. This kind of verification is necessary because the physiology, mindset and neurology of an elite player are completely different from those of even a semi-professional.

Close must have particularly persuasive powers, as the 16 players also agreed to give muscle biopsies and blood samples before and after a game. For the 36 hours leading up to a game, half of the players consumed a high-carbohydrate diet (6g carbs per kg bodyweight), including pre-match energy drinks and gels, and the other half consumed a low-carb diet (3g carbs per kg bodyweight). Through muscle biopsies Close could accurately measure glycogen levels before and after the game to determine the effect of each diet. The results were fascinating. Both groups exhibited similar performance results and similar glycogen levels *despite the difference in carb feeding*. Mosetter and Nemmer could be right – that pre-game gel is either not required or solely for the mind.

Half-time hit

The half-time whistle goes. In an ideal world, the support staff would swiftly determine individual players' glycogen levels and then administer bespoke solutions if required. A non-invasive ultrasound of the muscle (see box) has been mooted, but this technology has yet to catch on in Europe and so there's still a degree of

Measuring glycogen

MUSCLESOUND® IS A NEW, NON-INVASIVE METHOD OF MEASURING GLYCOGEN LEVELS ACCURATELY

Glycogen fuels high-intensity exercise, like accelerations and sprints, but determining the exact amount in your muscles comes down either to guesswork or the way more accurate but way more impractical biopsy. However, that could be set to change.

MuscleSound® is a non-invasive tool that measures glycogen levels via ultrasound. The athlete simply finishes a session, showers down, refuels and then waves a three-inch 'wand' over their major leg muscles. The wand is connected to a screen, which flashes up a score between 0 and 100, corresponding to the athlete's level of glycogen.

How does it work? It's all down to the principle that 1g of glycogen attracts 3g of water. 'When a muscle is full of glycogen, the ultrasound produces images of the connective tissue and fascia, but you can hardly see any muscle fibres,' says MuscleSound® founder and respected exercise physiologist, Iñigo San Millán. 'They are hidden by the water in the muscle.'

But after a long, hard, glycogen-depleting workout, the water disappears, exposing lots of muscle fibres. The more muscle fibres visible on the ultrasound, the less glycogen in the muscle. 'That's how we quantify it,' San Millán says. 'We developed software that reveals a glycogen score based on the ultrasound reading. It's like the fuel gauge on your car.'

To verify his invention, San Millán measured athletes' muscle glycogen content before and after exercise using both muscle biopsy and the new technology. The results, published in *Physician and Sports Medicine* in 2014, showed a near perfect correlation. He then commissioned a team of independent scientists from Appalachian State University's Human Performance Lab to replicate the study, and they also found a strong match. MuscleSound® was born.

On the MuscleSound® site, they evaluated an MLS midfielder's Muscle Energy Status (MES) throughout the season to determine trends. For example, the player's MES declined by 14 points after he had played four full games in a row. The player's MES improved by 15 points when he was benched for the next three games. Not great for his confidence, but perfect for his muscle energy stores!

In this increasingly data-heavy sport, regular MES readings can help to create personalised dietary and training strategies for each player in the squad. For example, consistently low scores are a warning that the player should consume more carbohydrates or reduce training intensity and duration; consistently high scores might suggest that the player should cut back on carbohydrates, which can turn into fat if glycogen levels remain high, or increase training intensity.

While many MLS clubs use MuscleSound®, the technology has yet to become popular in Europe. 'We hope to work with a Premier League club soon,' says MuscleSound®'s director of performance, John Ireland.

guesswork in even elite changing rooms. 'Many players will consume another gel at half-time and that's not a bad idea as studies show it can help you sustain more high-intensity runs,' says RB Salzburg's Asker Jeukendrup. 'Glucose levels also affect skill and cognitive performance, especially late in the game and that's when many goals are scored.'

An increasing number of energy gels also contain caffeine. And with good reason, says nutritionist James Morton. 'To this day, caffeine stands head and shoulders above every nutritional supplement [as a half-time hit]. It really works.'

Numerous studies show caffeine's ability to improve performance whether by increasing fat-burning and speed or lowering perception of effort. It does this by being a master of disguise. The neurotransmitter adenosine has an inhibitive effect – it promotes sleep and suppresses arousal. But when adenosine-lookalike caffeine is in town, nerve activity increases rather than slows down.

The pituitary gland in the brain senses this neural commotion, perceives it as an emergency and releases hormones that direct the adrenal glands to produce adrenaline, which stimulates the fight or flight response of increased heart rate and blood pressure. For a footballer, the result is an increased ability to repeatedly sprint at near maximum for more of the 90 minutes. (There's also evidence that caffeine accelerates glycogen synthesis after a match, but this remains equivocal.)

> Glucose levels also affect skill and cognitive performance, especially late in the game and that's when many goals are scored.
>
> **ASKER JEUKENDRUP** BARCELONA

How much caffeine a footballer should consume depends on the desired effect. If it's to cue the muscles for sprinting you'd need a significant amount, around 3–4mg caffeine per kg of bodyweight. That's around 200mg (two cups of coffee) for a 60kg footballer. If it's for endurance, 110–150mg would be sufficient. The espresso machine is a staple of every club but, says Southampton's performance nutritionist, Mike Naylor, every player should have a strategy.

'There are myriad proven benefits of caffeine,' he says, 'but you have to plan to maximise its positive effects. For instance, some players might consume a [caffeinated] gel 30 to 40 minutes before the game and then chew caffeine gum at half-time. This gets into your system within 10 minutes due to the chewing mechanism and how it's absorbed beneath your tongue.'

So proven are caffeine's performance benefits that WADA (World Anti-Doping Agency) used to prohibit its use above a certain concentration, but declassified it in 2004 because of its ubiquity in the food supply and the consequent difficulty in enforcing sanctions. The EU is also set to allow sports nutrition companies to explicitly state the performance benefits of caffeine in their caffeinated products.

There are caveats, however. Caffeine has been linked to gout, incontinence, headaches and indigestion. The Mayo Clinic in the US also presented a paper

▲ Coffee, uniquely decorated here with Jamie Vardy, has been shown to increase power output

showing how 250mg of caffeine resulted in elevated blood pressure for up to three hours, while a 2015 study by Italian cardiologist Dr Lucio Mos found that young adults who had been diagnosed with mild hypertension had four times the risk of suffering a heart attack if they drank four cups of coffee. The cardio-related impact of caffeine is something former Arsenal and England winger Paul Merson alludes to in his autobiography, *How Not to Be a Professional Footballer*, where he claimed to consume caffeinated tablets containing 'the equivalent of 10 cups of Starbucks. My heart started racing ... 30 minutes later that was still the case.'

Then there are evening games where teams might forego caffeine because its negative impact on sleep is felt to outweigh its positive effect on performance. 'This is particularly important if you have games in quick succession,' says Southampton's Mike Naylor, 'though if you did have caffeine, we'd try and counter it by simple touches like introducing milk before bed. There's evidence that tryptophan within the milk prepares you for sleep.

'And then, of course, some players don't feel they need caffeine,' continues Naylor. 'I get that. If you're walking out on to a pitch and there are 30,000 people screaming your name, do you need a caffeine shot or gum to feel alert? That'd be a big enough buzz for me. Also, some players refrain from caffeine because they want to feel more composed when they play.'

Post-match recovery

Once that final whistle goes, either to the tune of triumph or a chorus of contempt, it's into the changing room and time to refill the muscles with glycogen and repair and rebuild damaged tissue. 'I'd say like every club, I give the players a carbohydrate-and-protein shake after the game,' says RB Salzburg's Asker Jeukendrup. 'It gives me peace of mind that whatever the players do, they've had a certain amount of protein and carbohydrate.'

How much a player should consume is open to debate. There's evidence that glycogen in fast-twitch type IIa and type IIb muscle fibres, used heavily in high-intensity running, takes longer to replenish than glycogen in slow-twitch fibres used for walking and jogging. No one is certain why, but it could be down to the increased number of micro-tears in the muscle after intense exercise.

Jeukendrup aims for around 70g of carbs per player – maybe more for bigger players, less for smaller – and 30g protein. 'Timing is key, so, after the shake, we'll serve the players food while they're still in the changing room. Something nutritious and easy to consume with a spoon, like chicken risotto, which the chefs will have prepared and brought with them. After that, if they played at home, they'll probably go to the VIP area, spending time with family and friends and have another meal that's focused on carbohydrates and protein, all within around 90 minutes of the match finishing.'

If a team's playing away, that's where the team bus comes in. As we mentioned earlier, Nemmer serves the Liverpool players on their customised bus. Things aren't quite as sophisticated at Celtic, says the club's head of sports science, Jack Nayler. 'At the moment, our team bus isn't really set up how we want it,' he says. 'We have a microwave oven and fridges but we'd like a much more complete kitchen set-up at the back of the bus, where the chef could prep the food.

▼ Supplements at Southampton are based on the player and the demands of his game

> We use a term "performance backwards, not nutrition backwards". What does performance look like and work backwards from that.
>
> **MIKE NAYLOR** PERFORMANCE NUTRITIONIST, SOUTHAMPTON

'Still, we try our best. Take one of our last long away trips, we had a three-course meal for the players. As they got in, they were given homemade soup. Then a choice of curry, pasta, steak sandwich or fajita. And we gave them a dessert: sticky toffee pudding. We don't mind players having that kind of dessert as an occasional treat ... they are humans.'

Sticky toffee pudding might not be textbook sports nutrition but it's arguably better for the player than a fridge worth of berries. 'We refrain from high doses of antioxidants after a match and after training, too, actually,' says Peter Res, Ajax's head of nutrition. 'There's evidence that it impairs physical adaptation.'

Many follow Res's advice. One of the good things about free radicals (see page 107), which are produced in huge quantities when exercising hard, is that your body will adapt to them. They act as a stimulus. 'The problem is, when you deliver antioxidants in high doses, they sweep up the free radicals so you're cutting off the signal for the muscle to grow.' In fairness, Res is talking more about high doses of antioxidant pills than antioxidant-rich foods ... so Nayler can plant a strawberry on that toffee pudding if he sees fit.

Evidence-based supplementation

The issue of supplements extends beyond match day. Are they worth it? Some nutrition companies reportedly spend 90 per cent of their budget on marketing, and less than 10 per cent on research and development. It's where a football club's sports scientists and nutritionists really earn their stripes, by uncovering the merit within the marketing.

'One of my jobs is to keep up to date with worldwide journals and if we feel there's enough evidence to support a supplement's use, we'll try it,' explains Mike Naylor, part-time nutritionist at Southampton and the English Institute of Sport. 'But it must be relevant. We use a term "performance backwards, not nutrition backwards". What does performance look like and work backwards from that – what can we contribute from a nutritional standpoint? So James Ward-Prowse, for example, what are the demands of his game? Then we intervene nutritionally rather than coming in with a toolbox of ideas and just throwing everything at it.'

Beneath one of the monitors in the first-team canteen at Southampton's Staplewood training ground sits a collection of small plastic pots with initials on: SL (Shane Long), JWP (James Ward-Prowse)... Within each are individual daily supplements designed to improve performance, prevent injury and maintain ideal body composition. It's the same at Manchester City's training ground. One of the most popular supplements at Southampton, Manchester City and countless other clubs is beta-alanine. 'It's one that we feel contributes to repeated sprint performance throughout a match,' says Naylor.

Beta-alanine is a naturally occurring non-essential amino acid that is the building block of carnosine in the muscles. Carnosine is important because it counteracts the fatiguing effect of reduced pH in the muscles caused by the production of hydrogen ions in great numbers during high-intensity exercise. Carnosine increases the body's ability to buffer hydrogen ions and so maintains the pH balance within muscle cells. Daily doses of between 4.8 and 6.4g of beta-alanine have been shown to elevate muscle carnosine stores by 40–60 per cent in four weeks and by up to 80 per cent in 10 weeks. It is claimed that taking beta-alanine regularly can reduce fatigue over short and long sprints by 5–10 per cent.

The benefits at the elite end are clear. Players experience significant levels of fatigue as the game ticks by. Let's say James Ward-Prowse makes a 30m sprint at the start of the first half. He does the same with 10 minutes to go but is now 8 per cent slower. So a 30m sprint that might have taken 3.8 seconds at the start of the match will now take 4.1 seconds. In distance, this equates to around 2m.

Come the 83rd minute, if a fatigued Ward-Prowse is tracking a bursting run by a fresh central midfielder, his opponent has a 2m advantage, which is easily enough to evade the tackle and potentially create a goal-scoring chance. Theoretically, beta-alanine could help reduce this 2m deficit – and possibly prevent a goal being conceded.

Resisting fatigue and maintaining top-end speed can be the difference between success and failure. Take a look at the statistics. In the 2017–2018 Premier League season, 230 goals were scored between the 76th minute and full-time, the highest number for any 15-minute period in the entire game. That compares to the next best goal tally of 191 scored between the 61st and 75th minutes.

Of course, all kinds of factors other than beta-alanine levels contribute to late goals, both scoring and conceding, but a look at the teams with the best and worst goal difference in the last 15 minutes of matches gives some indication of which clubs are best able to maintain sprint endurance. For the record, and not surprisingly, in the 2017–2018 season eventual champions Manchester City topped the late-show charts, racking up the most goals, 25, and the best goal difference, plus 20, from the 76th minute onwards. Brighton, in contrast, had the worst goal difference of minus 11, primarily because they scored only two goals in the entire season after the 76th minute.

> We use creatine for players who want to put on muscle mass. It's also beneficial if you're an explosive player who's looking to execute more explosive sprints.
>
> **MIKE NAYLOR** PERFORMANCE NUTRITIONIST, SOUTHAMPTON

As with most performance-enhancing supplements, if you decide to use beta-alanine you must be aware of a side effect. A tingling sensation called paraesthesia can be caused by taking too high a dose at one time. This is down to sensitisation

▲ Sodium bicarbonate can boost repeated sprint speed … but there are colonic drawbacks!

of nociceptive neurons in the skin. Paraesthesia is harmless, though may be somewhat disconcerting. It can be avoided by splitting the daily dose into a recommended two to four servings.

Sodium bicarbonate is another buffering agent commonly ingested in professional football, with James Morton asserting that it was one of his go-to supplements during his time as Liverpool's nutritionist. Sodium bicarbonate – baking soda to you and me – elicits a similar effect to beta-alanine, but has been known to cause 'gastric distress'. Whether Gary Lineker had loaded with sodium bicarbonate before the 1990 World Cup match against Ireland when, in his words, he 'relaxed himself' in the middle of the pitch is unclear.

One of the most recent studies of sodium bicarbonate took place in that footballing hotbed otherwise known as the Faroe Islands, where Professor Magni Mohr and his research team showed that university players performed 70.3 per cent more high-speed running during the first five minutes of the match than they had before they started taking sodium bicarbonate. For the rest of the match the stats were pretty much identical, suggesting that the impact of the sodium bicarbonate, while significant, wore off quickly. The small study sample also makes those figures questionable and the fact that 30 per cent of the players enjoyed no benefit highlights the different effects that supplements can have on different people, a point picked up on by James Morton.

'Traditionally, you'd ingest sodium bicarbonate 90 minutes before exercise, though a paper published recently showed that everyone responds at a different rate. There's also evidence that when you perform a high-intensity warm-up, you can lose all the benefits because the buffering effects have been neutralised.' Mike Naylor says sodium bicarbonate is not on Southampton's list of supplements because of those potential gastric issues.

Creatine is another supplement that has been around the professional ranks for years (see box overleaf). 'In my experience, we use creatine for players who want to put on muscle mass,' says Morton. 'It's also beneficial if you're an explosive player who's looking to execute more explosive sprints. There's some evidence that creatine can enhance glycogen re-synthesis post-exercise, too, so if you're going through an intensive period like over Christmas, you might experiment with creatine.'

Creatine benefits

CREATINE IS ONE OF THE KEY SUPPLEMENTS ADMINISTERED BY FOOTBALL CLUBS

Why is creatine so prevalent in football (37% of professional Italian soccer players reported using creatine supplements according to a 2012 paper by Vincent Gouttebarge, Chief Medical Officer of FifPro, the World Players' Union)?

'Sprinting requires a high anaerobic component [generating energy in the absence of oxygen], which is covered by the creatine phosphate system,' says Jens Bangsbo, Professor of Exercise Physiology at the University of Copenhagen. That is exactly why many footballers take creatine.

The creatine phosphate system provides intense bursts of energy for activities less than 10 seconds in duration.

Every muscle contraction requires a molecule called adenosine triphosphate (ATP). When an ATP molecule combines with water, the last of the three phosphate groups splits apart and produces energy. This breakdown of ATP results in the muscle contraction and also reduces ATP to ADP (adenosine diphosphate) plus one single phosphate (Pi). You have limited stores of ATP, so they must be replenished.

And that's where creatine comes in, as phosphocreatine is then broken down by the enzyme creatine kinase into creatine and Pi. The energy released in the breakdown of these two molecules allows ADP and Pi to recombine and form ATP.

'By loading with creatine, there's an increase in pre-exercise phosphocreatine concentration,' explains Ted Munson, who formerly provided nutritional support to Hull City and now works as performance nutritionist at Science in Sport, who supply Liverpool. 'This enhances the capability of muscle to make ATP, thus allowing for more energy production and hence increased resistance to fatigue.'

Supplement the sun

The Christmas period is a time when another supplement comes to the fore: vitamin D. Often known as the sunshine vitamin, because it is synthesised in the skin through exposure to the sun's UV-B rays, vitamin D is widely associated with bone and teeth health but it also impacts muscular health. Few foods provide vitamin D, so as sunlight weakens during the winter months, there is a strong case for supplementation. In 2016, Public Health England announced new guidelines on vitamin D, suggesting most people 'should consider taking a supplement containing 10mcg of vitamin D, particularly during autumn and winter'.

Sports scientists have long been aware of the potential consequences of low vitamin D levels. 'There are seasonal trends with injuries and research shows a lack of vitamin D contributes to increased muscular injuries,' says Tony Strudwick, former head of performance at Manchester United. 'We found this was particularly relevant for players who'd come to the club from sunny climes – even in July and August, Manchester is cloudy!'

To combat the problem, Strudwick and his team took a threefold approach.

Creatine is an amino acid found in meat and fish and endogenously in the body. Wild game is the richest dietary source and free-range meats contain higher levels than factory-farmed equivalents. You don't find creatine in vegetables, which is why there's a particularly strong case for vegetarian footballers taking a supplement, available either in powder or capsule form.

'Whatever your diet, to produce any of the reported benefits of creatine, like repeated high-intensity efforts, muscle creatine content must first be increased,' says Munson. 'To do so, a creatine-loading strategy is used. There are two common strategies found to effectively maximise muscle creatine content. The first strategy is to consume a "loading" dose of 20g per day for five days as four 5g doses throughout the day, followed by a "maintenance" dose of 2–3g per day thereafter for 30 days. The second strategy is to consume the "maintenance" dose of 2–3g per day for 30 days.'

While many footballers swear by creatine, it does have certain drawbacks. Creatine loading leads to water retention, which can cause bloating and weight gain. Conversely, it can also induce dehydration, so continually topping up fluids is important. Creatine users have also reported unwanted side effects such as upset stomach, muscle cramps and dizziness.

▲ Creatine's use reportedly stretches to at least a third of players

'First, we gave players a high dose of vitamin D. Then we bought a sunbed and put it in the changing rooms. We also came off the training gas a little bit – almost a taper from October to December – before cranking it up from January onwards. We'd always do that with Sir Alex [Ferguson] and that's why we'd never smash the players in pre-season. If we weren't ready for that first game, it was never the end of the world. That's a hard thing for managers coming into the league to understand, that you need to manage a player's health and fitness across the whole season.'

The mechanism by which vitamin D affects musculoskeletal performance isn't fully understood, but several pathways appear to play a role. Multiple studies show that low vitamin D negatively affects handling, binding and storage of calcium in the muscle. Hence, one proposed role of vitamin D is to increase calcium accumulation in the muscle. Furthermore, phosphate imbalance, as induced by low vitamin D, was shown to cause muscle weakness, which can be reversed with vitamin D supplementation. Lastly, there's a direct effect of vitamin D on the muscle cells via the vitamin D receptor that stimulates new protein synthesis.

Arguably it's even more important for black players to supplement their vitamin D levels, as dark skin inhibits UV-B absorption because of the increased levels of melanin skin pigmentation.

Jelly legs

Vitamin D is a popular supplement; a touch more esoteric is gelatin, which is extracted from animal hides.

'We've improved recovery rates over the past couple of years by supplementing with gelatin,' explains Ajax's nutritionist, Peter Res. 'It's a powder and you dissolve it in hot water or in tea. Of course, normal gelatin forms a jelly but we use a hydrolysed version so it's less pudding-like.'

The theory goes that collagen within the gelatin is essentially the same 'material' as that found in tendons, ligaments and bones. When you break a bone or tear a ligament, the first bridges over the injury are comprised of collagen. 'The ingested collagen contains the same amino acids as the collagen in our body, so provides all the building blocks,' says Res. 'Specifically the amino acid proline gives the body a signal to step up collagen production.'

The protocol for an injured player is to take a gelatin supplement three times a day, says Res. There's also evidence that gelatin feeding provides a performance enhancement – namely more speed and power, both desirable football traits – by strengthening ligaments, tendons and cartilage.

Exercise itself increases collagen production, but combining exercise with gelatin supplementation has a particularly positive effect on collagen levels. A 2017 study led by Keith Baar, Professor of Molecular Exercise Physiology at the University of California, Davis, found that short periods of exercise with at least six hours of rest in between, to avoid muscle fatigue and damage, increased collagen production. When participants took 15g of gelatin an hour before the exercise activity – six minutes of skipping – the effect was to double the rate of collagen synthesis. This finding has connotations for injury prevention in football training. Players could take a gelatin supplement before a short session in the morning to boost their collagen levels and so increase their resilience to a more lengthy afternoon training session.

You snus, you win?

Caffeine, creatine, vitamin D – their acceptance as a supplemental strategy is as widespread as the rejection of another 'supplement': tobacco. A surprising number of modern footballers, including Dimitar Berbatov, Ashley Cole and Mesut Özil, have been known to enjoy an occasional cigarette, despite all that is known about the harmful effects of smoking on health and athletic performance.

However, a different form of nicotine has been in the news recently as a possible performance enhancer. At the 2016 European Championships, England's Jamie Vardy was photographed clutching a can of Red Bull (more caffeine!) and a pouch of powdered chewing tobacco known as snus. Vardy, who contributed 24 goals to Leicester's historic 2015–2016 Premier League-winning campaign, told reporters that he placed snus underneath his gums during games.

Snus has a long tradition in North American and Nordic countries – Ferrari's Finnish Formula One driver Kimi Räikkönen is often snapped chewing on a pouch. It is used by former smokers, like Vardy, to provide that nicotine hit, but has also been shown to enhance alertness, coordination and cognitive ability, improve aerobic performance and help weight control. Experts have calculated that it can boost athletic performance by up to 7 per cent and it has been on WADA's Monitoring Programme since 2013 to detect possible patterns of misuse. The sale of snus is also banned in every EU country except Sweden, so even if you were tempted to give it a try, you might find it hard to come by.

> At the end of the day, a professional footballer knows his or her body better than anyone. They will always have the final decision.
>
> **JAMES MORTON** NUTRITIONIST

Despite the best efforts of club nutritionists, it is ultimately down to individual players to decide what supplements they put into their bodies.

'We're scientists – it's our job to read through the literature, sift out the credible studies from ones that are less so and advise the players,' says James Morton. 'But at the end of the day, a professional footballer knows his or her body better than anyone. They will always have the final decision.'

You might question Morton's assertion that owners who shell out millions for their prized assets will leave the 'final decision' to the player, but there is an ethical argument here. It's one thing gently persuading a player to consume a tasty, natural meal high in protein if you're seeking muscle repair, it's quite another to insist on a regimented intake of synthetic supplements.

Then again, elite footballers are elite for a reason: they have a commitment to excellence that stems from their willingness to listen and learn in order to make progress. That's all down to the personality and mindset of the footballer, which are ultimately even more important at the highest level than physical attributes.

Fuelling through fixture congestion

WHAT SHOULD FOOTBALLERS EAT WHEN THE MATCHES ARE COMING THICK AND FAST?

Mayur Ranchordas is senior lecturer in sports nutrition and exercise physiology at Sheffield Hallam University. He's worked with many elite sportsmen and women, including players at Premier League clubs. Here, we present Ranchordas' nutritional strategies when there's less than 72 hours between games, as published in 2017 in the *Journal of the International Society of Sports Nutrition*.

PHASE	RATIONALE	PRACTICAL APPLICATION
Refuelling (post-match) / pre-loading (pre-match)	• A player should aim to consume around 6–10g per kg bodyweight carbs (e.g. 480–800g for an 80kg player) on the days where both muscle recovery and loading is needed (when there are 24–72 hours between games). This should be coupled with a reduction in training volume/intensity. • This is to be achieved through three to four main meals and regular carbohydrate snacking spaced out throughout the day. • Fuel intake should match the demands of energy expended. Players who have been an unused sub or only played part of a game do not require the same level of energy intake as players who played the whole game. Taking in more energy than required could lead to weight gain.	Carbohydrate sources to include as part of a nutritious meal: • Grains (quinoa, pasta, rice, noodles and couscous) • Starchy vegetables (potatoes), legumes (beans and lentils), fruits • Cereals (porridge, muesli) • Label foods appropriately to nudge players to increase carbohydrate portion for both match day -1 as well as post-match • Convenient food such as sweet potato wedges, chicken coated in breadcrumbs and chicken burritos served post-match can increase uptake.
Maintenance of repair and adaptation	• During intensive periods of competition a recommended strategy of 1.5g/kg protein per day (e.g. 120–160g for an 80kg player) should be sufficient to fully repair damaged muscle and stimulate soccer-specific adaptation. • Meals and snacks should be divided into six 20–25g protein servings over the day, eaten roughly three hours apart, to fully maximise protein-synthesis rates in the days between competition.	Protein sources containing 10g protein (add to carbohydrate sources for high-quality recovery meals): • 40g of cooked chicken, lean beef, lamb or pork • 300ml milk • Two small eggs • 30g of reduced-fat cheese • 120g tofu or soy meat • 50g canned tuna or salmon or grilled fish.

PHASE	RATIONALE	PRACTICAL APPLICATION
Rehydration	• Rehydration should begin as soon as exercise finishes. A player should aim to take in a volume that is approximately 150–200% of the estimated deficit to account for ongoing losses (e.g. urine output) with a rough guide of 1kg weight lost = 1.5l of fluid required. • Players should aim to replace the volume lost within two to four hours post-exercise over regular intervals to prevent the gastro-intestinal distress associated with large fluid intakes. • Key electrolytes need to be replaced – principally sodium – and this can be achieved either through electrolyte-containing drinks or consuming fluids with 'salty' foods. • Excessive alcohol consumption must be avoided as it is counterproductive to overall recovery goals.	Ultimately, fluid choices need to be palatable, suit the other recovery needs of the player, and be practical within their recovery environment: • Sports drinks containing electrolytes and carbohydrate • Milk-based drinks/supplements • Fruit juices • Cola drinks, tea and coffee could provide a valuable source of fluid and should not be totally avoided • Only have water if salty snacks are consumed at the same time.
Reduce inflammation and muscle soreness	• During intensive fixture congestion, antioxidants and anti-inflammatory food components or supplements can modulate the inflammatory reaction and may prove beneficial in the acute recovery phase. • Concentrated tart cherry juice and omega-3 fish oil supplements are two supplements that may accelerate recovery time. • It's vital to note that any form of antioxidant or anti-inflammatory supplement should be carefully dosed. Soccer-specific adaptations are triggered by the inflammatory and oxidation-reduction (redox) reactions occurring after a strenuous exercise stimulus.	Dietary sources of antioxidants include the majority of fruits and vegetables. High antioxidant-containing foods, for example: • Blueberries, prunes, sprouts, broccoli, raspberries, sweet cherries Dietary sources of omega-3: • Oily fish, beans, flax seeds, walnuts.

7

BRAIN TRAINING

Go on YouTube and search for a series of videos in which scientists search for the cognitive secrets behind Cristiano Ronaldo's success. The calm, soothing narrator talks us through Ronaldo being fitted with a state-of-the-art tracker that measures eye movement to monitor what the Portuguese star focuses on when he plays. With his swift feet, Ronaldo aims to keep the ball away from the presenter/defender. In eight seconds, he executes a staggering 13 moves – spins, step-overs – leaving the opposition bamboozled.

In slow motion, his leg speed and balance are even more incredible; it's a poetic piece of physicality. But the true orchestrator is the brain. So just what is going on in his mind? More specifically in this experiment, what was Ronaldo looking at? As it transpires, almost everything but the ball. The 'gaze data' showed that he was focused on the defender's feet, hips and knees, looking for subtle clues to reveal his opponent's next move. He was also surveying the space beyond the defender, searching for a way to escape. Which he did – with ease. As for the defender, he was transfixed on the ball, meaning he could only react rather than predict.

Inner GPS

Vosse de Boode, head of sports science at Ajax, has worked on eye-tracking projects for several seasons and, as highlighted in chapter 2, feels technologies like these work particularly well in specific match situations.

◀ Cristiano Ronaldo simply sees space and movement that most players don't

'One of our goalkeepers claimed that they didn't see the ball when a penalty was taken,' de Boode explained from the team's miCoach Performance Centre. 'So we asked: did he really not see it, or did he see it and make a decision based on how he processed the information? With the eye-tracker, we examined what our goalkeepers focused on right before the shot, and noticed that those who made an early, incorrect decision were looking directly at the feet of the opponent. Other, more successful goalkeepers were waiting a fraction longer for a specific piece of information – body shape, angle of approach… That then transferred to the physical movement. Essentially, the faster you can jump sideways, the longer you can wait before making that decision. We're talking milliseconds, but that's all it takes.'

De Boode and her team share this insight with the goalkeeping coaches. 'You can then analyse why the goalkeeper made that decision – was he not sure about the technical side? Was it something that the coach didn't explain clearly enough? It's often more a starting point for discussion but the team can also work on specific areas, like getting the keeper to focus on the correct part of the penalty-taker's body.'

Dr Kurt Mosetter, the founder of Myoreflex Therapy (see page 109), who has worked with a host of German teams including Bayern Munich, says this ties in with the inner GPS, a concept that explains how the brain creates a map of the space surrounding us to help us navigate our way through a complex environment, in this case a football pitch. In 2014, scientists John O'Keefe, May-Britt Moser and Edvard Moser received the Nobel Prize in Physiology for their research in this area, including the discovery of 'grid cells' in the brain that generate a coordinate system.

'The inner GPS is also important from an injury perspective,' Mosetter explains. 'If a player has an accident – maybe a knee pain, hernia or cruciate – the information from the GPS changes. So if we immobilise that area for even as little as a week, the inner GPS will establish a new pattern and the brain's ability to predict elite footballing movements has been impaired. So you've now got two injuries to treat.'

It's one of the reasons why early mobility post-injury is recommended, and why the old rehab acronym of RICE (Rest, Ice, Compression and Elevation) has been replaced by POLICE (Protection, Optimal Loading, Ice, Compression and Elevation) (see page 95).

'We undertook eye-tracking with RB Leipzig players,' Mosetter continues. 'With that information, we then covered up the players' better eye [we all have one eye that works better than the other], which improved their information-processing channel through their worse eye in a very short time. The inner GPS had "realigned" itself, resulting in faster decision making and improved awareness of orientation in space.'

It's these sort of response-reaction results that have attracted the likes of Monaco, Wolverhampton Wanderers, Tottenham Hotspur, the England rugby team and, recently, Premier League referees to South African sports scientist and eye expert Dr Sherylle Calder. Dr Calder has designed a structured on-field programme and computer software to improve response speed, concentration, spatial awareness and decision making.

Calder insists her methods aren't designed to improve eyesight but to broaden perception. One reason she feels this is necessary is to counteract the negative effects of fixing your gaze on a smartphone, a habit she believes has led to a marked deterioration in footballers' visual awareness over the past six or seven years.

Like Vosse de Boode's efforts at Ajax, much of Calder's work is about developing skills to see things early; to give yourself more time to make the appropriate decision. 'This drill is one I'd use with teams and is designed to encourage players to think and react rapidly,' she explains by way of an example. 'You need 12 players – six each side, one with red bibs, the other with blue – and an area 20–30m square. You score a point for eight consecutive passes. Both teams have a ball, passing the ball in a "one-touch, two-touch" sequence. The passer must remind the receiver how many touches they've had. This progresses to both teams having two balls each, passing the ball freely, before moving on to one-touch, two-touch again. Again, the passer reminds the receiver how many touches they've had.

'You then have one ball shared between both teams, and red can only pass to blue and vice versa. The passer must remind the receiver what colour they must pass to next. The drill is finished off with two balls shared between the teams. Again, red can only pass to blue and vice versa. The passer must remind the receiver what colour they must pass to next.'

Fluid intelligence

'Look at players like Xavi and Iniesta, they have a really high cognitive ability to pick up patterns of play,' explains Malcolm Frame, Southampton's long-time sports psychologist. 'It's called fluid intelligence.'

▶ Andrés Iniesta (left) has been scientifically proven to have a high cognitive ability to pick up patterns of play

For an example of Andrés Iniesta's 'fluid intelligence', you only have to watch him almost single-handedly tear Real Madrid apart in 2015 – the diminutive Barcelona midfielder scoring one and setting up two before being substituted with his team 4–0 up. Iniesta left the pitch to a rare standing ovation from the opposition fans at the Bernabéu, not just for the goal and the assists but the flicks, the passes, the movement, seeing space where no one else could.

According to a 2012 study by researchers from the Karolinska Institute in Stockholm, fluid intelligence can be measured. A sample of 57 footballers from the top three divisions in Sweden took a standard test to assess their 'executive function' – their ability to be creative on the spot, find new solutions to problems and learn from what has not worked before. The study produced two surprising findings. First, that the professional footballers scored more highly than the general population, with the players in the higher divisions outperforming those in the lower. Second, that the players with the higher scores tended to rack up more goals and assists over the following two seasons, so the scores could be taken as a predictor of future success.

'The researchers then did the test with Xavi and Iniesta,' says Frame. 'Both scored very highly in it, because it's all about shapes and patterns.' In fact, Iniesta was in the top 0.1 per cent for 'design fluency' – problem-solving behaviour exhibited in fluency in generating visual patterns and creativity in drawing new designs. They also scored highly in what neuroscientists call 'inhibition' – the ability to alter your approach in a way that makes it easier to complete a task. 'The key is learning from this,' adds Frame. 'How can we train our players to improve their executive function?'

▼ The Karolinska Institute discovered that players with a high executive function scored more goals and assists than their contemporaries

▲ Executive-function games like Tetris 'might' be able to improve a footballer's skill-set

One method proposed uses a variety of computer games (supporting the theory that some games – Tetris, for example – can improve a footballer's ability on the pitch). These executive-function games could even be used by academy coaches to help them select new recruits.

When the automatic becomes controlled

Executing a skill, pass or decision at speed requires a subconscious awareness, the ability to perform on 'auto-pilot'. It's perhaps easier to appreciate this point by looking at what happens when it *doesn't* work, and few teams have provided more examples than the England football team.

Dan Abrahams is a sports psychologist who has worked with clubs including Derby County, West Ham, QPR and Fulham. He has also worked with the England rugby team under Eddie Jones. He cites performance anxiety and the cascade of negative biological changes this entails as one of the key reasons why 1966 remains the one and only time the Three Lions have opened their trophy cabinet door (despite a valiant effort in 2018).

'Scientific research would say that what happens when a team like England underperforms is that the automatic processes become controlled,' Abrahams explains. 'One of the more overt cases is the Iceland game [at the 2016 European Championships]. I have it on good authority that when they went 2–1 down, there was an overwhelming feeling of embarrassment. What happens then is that automatic processes in the brain, which are wired deep in the limbic system, suddenly become a frontal-lobe function. In other words, the automatic becomes controlled. Many accuse England players of not trying hard enough, but, actually, often they're trying *too* hard.'

A further example of the automatic becoming controlled came in the 2010 World Cup when England played out a tedious 0–0 draw with Algeria and, particularly, in the performance of talisman Wayne Rooney. That season, Rooney scored 34 goals in all competitions for Manchester United. It won him the PFA Players' Player of the Year and FWA Footballer of the Year awards.

Rooney went to the South Africa World Cup with the expectations of a country upon his shoulders. They proved too much as the simple became complex, the automatic became controlled. The *Guardian*'s Paul Hayward summed up his performance: 'Whatever Rooney achieves in South Africa will require an inner victory over the frailties holding him back because it all looks woefully like labour, as if he were a pianist whose timing has temporarily deserted him.'

Performance anxiety also affects a player's hormonal profile. 'Their confidence drops and they release the stress hormone cortisol,' says Abrahams. 'Behaviourally, that leads to tunnel vision. They no longer see the space, teammates or movement of teammates or opposition so quickly. They lose their anticipation and won't create space to receive the ball. You can see it through body language and more objective measurements like GPS data.'

It's not solely an England problem, of course. Abrahams describes the Brazilian team who were thrashed 7–1 by Germany in the 2014 World Cup semi-final as saturated in cortisol. But the million-dollar question is how can England, or any team or individual player, counter this performance anxiety or, preferably, prevent it happening in the first place?

▲ At the 2010 World Cup England were beaten by Germany ... and huge levels of cortisol

Picture it, speak it

'Self-talk and visualisation are key tools in the players' armoury,' says Abrahams. As the name suggests, self-talk involves the player developing an inner voice that encourages peak performance. This can be action-based, so 'keep sprinting beyond the defender' or simply 'keep working hard'. Or it can be attitude-based: 'keep playing with freedom' or 'keep focusing'. (As an aside, staying focused is also a key psychological skill of any elite sportsperson – so much so that legendary Northern Irish goalkeeper Pat Jennings used to come off the pitch with a headache after concentrating so much.)

Visualisation is creating mental imagery of a specific action without any corresponding motor output. In essence, you imagine your physical performance without actually physically moving. It's a proven technique supported by a significant amount of anecdotal feedback. Javelin thrower Steve Backley is renowned for winning silver at the 1996 Atlanta Olympics despite being unable to throw a javelin during the build-up after rupturing his Achilles. Backley replaced physical training with visualisation, working with hypnotist Paul McKenna. 'He helped me visualise the perfect throw, down to the time it took to throw and land. I saw it in 3D,' Backley said at the time. 'My daily mantra became "see it, feel it, trust it".'

Wayne Rooney reportedly asked his kit man which strip the team would be wearing in an upcoming game so he could more accurately picture himself scoring, while Joe Hart visualised himself having the game of a lifetime when preparing for Manchester City's 1–0 defeat to Barcelona in March 2015. Although City lost, Hart played superbly, making a record number of saves for an English goalkeeper in a Champions League match. And it's all supported by science, according to Ian Robertson, Professor of Psychology at Trinity College, Dublin. 'If you put athletes through an MRI scan while they are visualising performance, almost all the parts of the brain that are active when you're actually doing it are active when you're imagining it,' he says. 'It's only the final pathways tied in with sending signals to the muscle that aren't active.'

Dan Abrahams advocates melding self-talk and visualisation, to create an all-round stronger, more practical tool. 'I'll sit down with players and ask them to tell

▼ Joe Hart is a big fan of visualisation. It paid off in 2015 when Man City beat Barça 1–0

me about themselves when they were at their best. What did that look like? What did that feel like? We'll break that down into a few action words: "Well, I was dominant and relentless." They'll picture that in their minds – dominant and relentless. "Yeah, moving on my toes"; "I'm alert and lively". Eliciting good, positive action-based words that they can use on the pitch.

'Then we might have some fun with it and relate those qualities to an animal or a model player. So players will go out and be a predatory lion or a speedy greyhound. Or a sharp, relentless Lionel Messi or a dominant, confident Vincent Kompany.

'That's the way a human works best when talking about performance. If I have a player low on confidence, I want them to picture themselves at their best, boil that down to key words that they can access and act out on the pitch, and then make that image richer, more vivid and more football-specific.'

BLR Ronaldo

There are sceptics – British football has been slow to embrace sports psychology and clearly self-talk and imagery can't transform a Sunday morning Hackney Marsh hacker into Thomas Müller – but nearly all professional footballers will use these techniques in some form. Back to Abrahams: 'I've been working with Yannick Bolasie [Everton] for a few seasons and his key words and imagery through a game are BLR Ronaldo. That stands for "brave, lively, relentless Ronaldo". When it comes to big games, the brain naturally wants to feed thoughts like "I've got to play well as it's on TV" or "I want to score goals". That's when cortisol is released.

'So we've worked on his BLR Ronaldo. What does that mean in terms of body language? It means *brave* runs, taking people on; staying *lively* and on his toes at all times, even if he's not on the ball. *Relentless* might mean making repeated runs into the 18-yard box to give himself a chance of scoring. Ronaldo is a great cue word for him because it snaps him back. What would Ronaldo do now? How would he react? Basically, players need to intellectualise the self-talk and visualisation for their role and responsibilities. They then have a blueprint to take on to the pitch in times of adversity. Say it, do it, be it.'

Visualisation is a highly individual technique and the way it's taught by practitioners varies, too. Dave Collins is a performance psychologist of over 25 years' experience and currently works with 'a number of Premier League players, but

If I have a player low on confidence, I want them to picture themselves at their best, boil that down to key words that they can access and act out on the pitch, and then make that image richer, more vivid and more football-specific.

DAN ABRAHAMS SPORTS PSYCHOLOGIST

▲ Yannick Bolasie (right) recites BLR Ronaldo. 'It stands for brave, lively, relentless Ronaldo,' says psychologist Dan Abrahams

I can't say who'. He's also a fan of visualisation. 'There's the old idea of visualisation that's all about relaxing in a darkened room, closing your eyes and running through things in your head,' he says. 'And, yes, that still has a place. But it's also beneficial to use mental imagery in a similar physical state to when you'd need to execute a skill. When I worked in judo, for example, we'd have fighters doing puke-inducing anaerobic sessions and when they would walk back in for recovery, I'd have them image-fighting their toughest opponents. Because when they're fighting their toughest opponents, they're going to be on the edge of puking up. That makes it more vivid and makes the transfer better.'

Collins calls visualisation the Swiss Army knife of sports psychology because 'there are so many ways you can use it'. And that list is growing and potentially becoming more vivid thanks to virtual reality (VR). Because of motion-sickness issues, it's still early days for the application of VR in sport. But it's happening. When I visited Southampton, head of performance science, Mo Gimpel, showed me his tidy-ish desk that was bulging under the weight of a Mac, PC, piles of paper and a Google Cardboard viewer into which you can fit a smartphone for using VR apps. 'It cost a

Virtual imagery, real progression

MI-HIEPA SPORTS ARE BRINGING VIRTUAL REALITY TO THE WORLD'S BEST CLUBS, INCLUDING MANCHESTER UNITED

At the inaugural Soccer Science conference at Bristol City's Ashton Gate in June 2018, one of the most attention-grabbing features was the virtual reality (VR) suite demonstrated by Manchester start-up Mi-Hiepa Sports.

'There is nothing else like this on the market,' co-founder Adam Dickinson told me. 'It can be used for tactical reasons, improving decision making, returning from injury... You can even use it to scout players.'

Mi-Hiepa takes you into the virtual footballing world via an HTC Vive headset, customised boots and RDX shin pads, featuring sensors, all connected to powerful software. 'We use auditory and visual signals to give you the feeling of kicking a ball,' explains Dickinson. 'Most VR systems use haptic technology [which recreates touch sensations] but, in our opinion, that's not as real as this technology. Imagine you're kicking a ball – you'd need a haptic device all the way around your foot. If the ball just brushes the top of your foot, or if you back-heel it, then that's going to feel weird. We've been careful to create something incredibly immersive that is far beyond what people would think possible.'

The playing area is 10m square, so you've got plenty of space to move around, but it's what's in front of you that really impresses. I gave it a go, slipped the VR headset on, looked at my feet and noticed my left football boot was yellow and red, the right pink and blue. 'When the cannons fire the balls at you, they'll correspond with one of those colours,' Dickinson said. 'Simply strike the ball with that colour part of the foot. If they're silver balls, you can strike with either foot; if they're grey, you must avoid them.'

tenner, I think,' Gimpel said. 'We're looking into this area to see how we can use it.'

With its interactive, immersive and multi-sensory qualities, it doesn't take a leap of imagination to see how VR can benefit football. If you're Arsenal full-back Héctor Bellerín and you're playing Chelsea the following weekend, you might want to immerse yourself in 3D video clips of Marcos Alonso to study his movement, whether he cuts inside or out, his favourite tricks.

It'll be an easy buy-in, too, as the youngsters coming through the academies are digital natives, in contrast to older generations of digital migrants. Gimpel says Southampton are in talks with Microsoft about using their HoloLens units, while Stoke City are currently using VR with their goalkeepers after signing a deal with Dutch firm Beyond Sports. The company also work with PSV Eindhoven, AZ Alkmaar and the Dutch national team.

Beyond Sports have recently created an 'analysis suite' for Arsenal at their training centre in Hertfordshire, decked out with televisions, VR units and a small pitch. The north London club were apparently convinced by Beyond Sports' claim that their units helped Ajax academy players improve their decision making by

Cue me dancing, jumping and kicking around Ashton Gate. 'You scored 66 points,' Dickinson commented. 'Not bad.' As it transpired, not that good. England goalkeeper Jack Butland got 186.

Still, I could see how the suite could refine technique and improve spatial awareness. Dickinson then switched games, sending me into a virtual stadium as part of a back four. 'It's set up so that Raheem Sterling scores. Just see how quick he is.' I catch him out of the corner of my eye. He's damn quick. I'm then transported from ground level to the top of the stand to run through the same move but visualising it from a different angle.

'You can see how coaches can run through real-life situations after the match and more vividly show the players where things went wrong and what they could have done instead. Most clubs want it to be individualised, which we can do quite easily. We just take the actual footage of the goal, put it through our system and deliver it to the club to work on.'

It's impressive stuff and opens up a whole new world of tactical possibilities. Four club defenders who might be on international duty for different national teams, for instance, can take a VR unit with them and virtually link up with the other three defenders to practise real-life drills they might have worked on before they went away, each seeing where the other is and how they should individually and collectively respond to certain scenarios.

Because it's non-contact, the system can also be used to help injured players keep up their ball skills while they recover, and there's also a scouting option to test prospective signings without their having to travel to you.

I'm no professional footballer but I can see how the system could complement existing physical, skill-based and tactical training. And, according to Dickinson, the high frame rate and the fact you can see your feet, mean there is no problem with motion sickness. After my test run, thankfully I can confirm this.

20%. However, Özil and his colleagues are reportedly ambivalent, with motion sickness still an issue.

Train the brain to ignore the pain

Robin Thorpe is Manchester United's head of recovery and regeneration. When he spoke at the Future of Football Medicine conference in May 2017, his closing slide showed a jigsaw piece that featured the sentence: 'Mental fatigue may be the missing piece in the puzzle…'

Professor Samuele Marcora would agree. Director of research in the School of Sport and Exercise Sciences at the University of Kent, Marcora is a great believer that fatigue is a mental phenomenon not a physical one and has spent years proving his psychobiological model of fatigue. 'When effort is perceived as maximal or when effort required eclipses the amount of effort you're willing to exert, then you stop,' explains Marcora. 'In essence, fatigue is an increase in time of perception of effort.' For Marcora, the act of stopping is a conscious decision rather than a mechanical response.

Marcora examined the impact of mental fatigue on footballing performance by conducting two separate studies of a group of trained footballers, to measure the effect of mental stress on physical performance and on technical performance. Both studies also examined perception of fatigue after a cognitively demanding task.

The first study saw the 12 players divided into two groups: half read a magazine for 30 minutes while the other half performed the Stroop test of executive function. This is a common neuropsychological test where the subject is shown a list of words that spell different colours but with letters of a different colour. For example, the word 'red' may be shown in blue letters. The subject has to call out the colour of the ink rather than read out the word and vice versa. It sounds simple, but requires concentration. Both groups then performed the Yo-Yo intermittent recovery test (see page 45) and their performance heart rate and Rate of Perceived Exertion (RPE) – essentially, a fatigue self-assessment – were recorded at the end. As Marcora predicted, the Stroop group performed worse in the Yo-Yo test and reported a higher RPE despite having a similar heart rate to the magazine group. Why? Because they were more mentally fatigued.

Marcora then performed another study with the footballers to gauge the impact of stress on technical performance. This time the Yo-Yo test was replaced by a passing and shooting test. Again, those who had performed the mentally challenging Stroop test fared worse than the 'relaxed' magazine group.

Having observed the negative impact of mental fatigue on physical performance, Marcora went on to experiment with using mental exertion as a way of achieving training adaptations, similarly to how one would use physical exertion. He homed in on the anterior cingulate cortex (ACC), a section of the brain associated with effort perception, reasoning that if we can strengthen the ACC through 'brain endurance training' then our perception of effort will be reduced and our physical performance improved.

To test this theory, Marcora put two groups of soldiers through 12 weeks of riding an exercise bike three times a week, with one of the groups carrying out a mentally demanding but unengaging task during the sessions and the other group doing nothing but pedal. When he conducted a 'time to exhaustion' test without accompanying mental task at the end of the 12 weeks, he found that the soldiers who had been exerting their brains during the sessions had improved their fitness by three times as much as the other group.

Marcora's model has detractors, who accuse it of being overly simplistic and too biased towards motivation, but many agree that an organ that consumes 20 per cent of a human's daily energy supply – the brain – can be trained to make the body perform for longer.

Could Messi be even better with a daily 15-minute dose of a dull, but mentally taxing task? Debatable. But Marcora's model is all about improvement; about making a good footballer great. Which begs the question: what does make a player great?

What forges greatness?

In 2016, Gordon Taylor, chief executive of the Professional Footballers' Association, said that of the boys who make it into the elite scholarship programme at 16, five out of six are not playing professional football at 21. (A similar programme for young female footballers was announced in 2017.) Is this rate of wastage unexpected? Perhaps not. But by any standards, these are talented individuals who've likely racked up the famous 10,000 hours of deliberate practice. They can sprint as fast as Walcott, jump as high as De Gea and cover the same ground as Kanté. So what is it that separates the few who make it from the many who don't?

'There have been numerous studies on what makes a top sportsman,' says performance psychologist Dave Collins. 'These have identified certain qualities, including: self-regulation – the ability to handle pressure; use of imagery; realistic performance attribution – maintaining a balanced viewpoint whether you win or lose. Use of social support is a big one, too. Players who reach the top seek out and use the right advice. There's also the ability to set goals and to praise yourself when you've done something right. And it's important to be able to elicit quality practice all the time, to go out and say "I'm going to work on this at a level of intensity and apply myself."'

Self-control and self-regulation are arguably the two most important factors in cultivating footballing excellence. It's a topic researchers Tynke Toering and Geir Jordet investigated in 2014 by comparing the self-control scores from a questionnaire given to teams in Norway's top two leagues. A total of 639 players assessed their self-control across 13 different factors, from 'I have a hard time breaking bad habits' to 'I am able to work effectively on long-term goals', rating themselves on a five-point scale from 'not at all' to 'very much'. Essentially, the higher-performing players had higher scores for willpower in the face of distraction and temptation.

The ability to learn for yourself is crucial, according to sports psychologist Dan Abrahams. 'You need to be able to take on messages and create a story in your mind about how you can use those messages,' he says. That self-reliance results in a strong work ethic. Cristiano Ronaldo was forever seeking out Sir Alex Ferguson's right-hand man René Meulensteen for extra drills and sessions, while further anecdotes abound of how hard a young Ronaldo worked at Sporting Lisbon. 'There's one story where he appeared to be running a lot slower than normal and didn't seem to be performing as well,' Abrahams recalls. 'The coaches asked if he was OK. He lifted up his trousers and he'd slipped in weights beneath his socks because he wanted to work on his leg strength. It's incredible commitment and, of course, Ronaldo had experienced trauma. Talent likes trauma…'

Arguably the greatest player in the world underwent surgery on a heart condition when just 15, was expelled from school and lost his dad to alcoholism when he was 20. 'It doesn't need to be that severe,' says Abrahams, 'even a small amount of adversity can mould an elite athlete.'

Team dynamics and social architects

Individual greatness is one thing, but what makes a great team? It's the eternal question. But really, what unites the 1970 Brazilian side, Cruyff's Ajax and Guardiola's tiki-taka-ing Catalans? Steve Peters is a psychiatrist who's best known for his work with Team Sky and British Cycling, but he has also worked with England and Liverpool. His groundbreaking book *The Chimp Paradox* is based on the premise that for survival we need to harness our impulsive and reactive emotions, because if they're not managed properly, they can have a catastrophic impact.

He works one on one with players and with backroom staff. 'When I'm working with different football clubs, it's clear that each one has its own culture,' he says from his Peak District home. 'There are behaviours that are acceptable and not acceptable, and there are thinking and beliefs behind those values. These formulate a culture. When you're a player at a club where the beliefs, behaviours and values are very different from your own, unless you can blend in it's unlikely you'll do well there.'

This could apply to players coming into the club or players expected to play a different role from the one they're used to. 'Take a footballer on the pitch who's asked to do something different,' Peters says. 'We might have to work differently with his mind because he might not agree with what he's been asked to do, which means he has to come to terms with that. He has to recognise he's part of a team, not an individual. Sometimes you have to elicit conflict out of someone because they don't recognise that. My job is to ask questions, find out behaviours, work out why the player's unhappy and put it back again. They might say I've got it wrong so then we try again.

'I don't tell people whether they're right or wrong. I observe and give them my opinion.'

> When you're a player at a club where the beliefs, behaviours and values are very different from your own, unless you can blend in it's unlikely you'll do well there.
>
> **STEVE PETERS** PSYCHIATRIST

Peters' work is particularly meaningful and influential when he is able to create a ripple effect through a whole team. 'When observing a team, you might note one player who has a greater impact on the team than another,' he says. 'We call these people social architects – someone who has a heavy influence on a culture, whether it's in business, sport or even a group of friends. There will be leaders who set the pace and culture.

'So, if everyone's happy for you to do so, you might go in and work with the social architects. They aren't always the obvious people. For example, sometimes quiet individuals have an important calming effect on the team. A social architect is anyone who has a big influence on the way that the team think.

'Observation might show you which factors are affecting performance and social architects might be one area. You might have five social architects pulling in different directions so that the team lacks cohesion.'

▶ Eric Cantona was one of Sir Alex Ferguson's key social architects

One social architect who had a massive impact on a team is mercurial Frenchman Eric Cantona. In Professor Damian Hughes' book *How to Think Like Sir Alex Ferguson*, he reflects on Cantona's first training session at Manchester United after moving from Leeds United:

> *At the end of the session, as his teammates were vanishing from the pitch, the Frenchman approached his manager and asked if he could have the assistance of two players.*
>
> *'What for?' Ferguson asked.*
>
> *'To practise,' he replied.*
>
> *'That took me aback,' admitted Ferguson. 'It was not exactly a standard request but I was naturally delighted to accede to Eric's wishes.'*
>
> *Meanwhile, the players who'd gone inside were realising that Cantona had not come back in and began to explore why. 'At the end of training the next day, several of them hung around to join in the practice with Eric and it soon became an integral part of my regime,' Ferguson explained.*

Delivering the message

A great team, of course, requires a great manager. The best ones, says performance psychologist Dave Collins, are exceptional psychologists. They also manage to blend the stable and the unpredictable (Gary Neville and Cantona, for example) for 'magical outcomes'. And they're great communicators, too. Much of Collins' work with teams centres on communication, and like many sports psychologists, he has found it's often more effective to work directly with the communicators – the managers and support staff – than the players. They become his voice.

'A clever manager in any sport will use a variety of ways to communicate. He will work through the media to send messages to his players; he will work through the fans and through teammates. If you were to monitor media outlets of one of our top managers for two or three weeks, you would see them doing this. They'll do it with individual players, the whole team, units of the team...'

Here's an example. After winning one in eight games between 9 September and 14 October 2017, Liverpool manager Jürgen Klopp became the focus of the media spotlight, newspapers and pundits questioning his ability to turn the Anfield side into a league-challenging outfit. But Klopp held his nerve... After a 1–1 draw with Burnley on 16 September: 'We and all the other teams will have moments in the season that are a little bit more difficult to go through. That's completely normal.' After another 1–1 draw, with Newcastle on 1 October: 'The big challenge for us is to stay confident

▼ After one win in eight, the Liverpool players got the message with a 7–0 win over Maribor

and play our football. I am ready for this and the boys are ready for this.' After drawing 0–0 with Manchester United on 14 October: 'I think United came here for the point and they got it. We wanted three points and didn't get them. For me today, one team who can become champion this year was in our stadium and is not a world apart from us.' Despite the growing pressure, Klopp remained upbeat, positive, telling his team it would come good if they kept working hard.

A couple of days after the Manchester United match, the team flew to Slovenia to play Maribor in the Champions League. Maribor are minnows compared to Liverpool but were unbeaten in domestic competition and had held Chelsea at home back in 2014. Another draw seemed the most likely result. Any dissension in the ranks would come to the fore under the Ljudski Vrt stadium floodlights. It didn't happen. The Reds won 7–0, the biggest away win by any English team in the history of the Champions League and European Cup. 'Players need that clarity and, often, reassurance,' Collins says. 'Things get difficult when individual players are receiving mixed messages about what's expected of them or being lambasted. They go in different directions.'

Which is what happened to Harry Redknapp, who in his most recent jobs gained a reputation for blaming everyone but himself. This from January 2015 with QPR second to bottom: 'I got the maximum out of my team again. I can't do any more than that.' Two weeks later, he'd resigned, citing knee issues, though many felt he jumped before being pushed (which obviously wouldn't have helped his bad knee!). His last job was at Birmingham City. After losing 2–1 to Burton in August 2017: 'Nobody can say, "OK, lads, I'll sprinkle stardust on you, you're a new lot of boys this year, forget that you ruined [former manager] Zola last year, this year we're going to be different".' The following month, just 13 games into the new season, he was sacked.

Of course, the messages managers deliver aren't always for the team. Take José Mourinho. In October 2017, with Manchester United still unbeaten, Mourinho was interviewed on French TV programme *Telefoot* and responded to the question of whether he might one day manage PSG by revealing his son had taken the club to his heart. Asked why, the United manager replied: 'Because at the moment in Paris there is something special. Magic, quality, youth, it's fantastic.' It seemed an incongruous statement, especially for a man then riding high near the top of the table. But Mourinho's not stupid. His contract had 18 months to go. He knew his worth. He wasn't communicating with the interviewer, he was talking to chief executive Ed Woodward and the team upstairs at Manchester United: 'Extend my contract, increase my money … or I'll be heading to Paris.'

Learning from the class of '92

Would Mourinho's illustrious predecessor Sir Alex Ferguson have used that strategy? Probably not. And was it coincidence that soon after Mourinho's interview, the team suffered a drop in form that included a 2–1 defeat to Premier League newcomers

Huddersfield? A man who'd know better than most how Ferguson would have reacted is sports psychologist Bill Beswick, who worked with Fergie in the golden years.

'When I moved from Derby to Manchester United, I found an amazing collective winning mindset. They trained harder than any team I've ever seen before, the personalities were incredible: Keane, Giggs, the Nevilles, Beckham… They were the strongest I'd ever come across. And they had a belief honed by Alex Ferguson that was so powerful and targeted to winning that it was a real case study for me of what creates an elite environment in a football club and how a psychologist can help that.'

> Winning after winning is the hardest thing in sport. Complacency is just around the corner.
>
> **ALEX FERGUSON**

The day after speaking to me from his Cheshire home, Beswick was heading south. He's contracted to Bristol City for 25 days a year, mentoring the Robins' manager, Lee Johnson, and his charges. City were fourth in the Championship at the time and challenging for Premier League status. The last time they had played in England's upper echelon was 1980.

Beswick uses his 25 years as a sports psychologist and, specifically, what he learned under Ferguson when the Class of '92 et al dominated English and, occasionally, European football. 'Many tend to look at performance as physical, technical and tactical, but the good managers realise those three are driven by the mindset of the individual players and, in a team sport, the mindset of the team. And United had that mindset in bucketloads.'

Many managers call in a sports psychologist as a last resort. Ferguson, on the other hand, brought Beswick in to maintain the winning mentality. 'What people don't realise is that there are as many psychological problems when you're successful as when you're not. They feel more obvious when you're losing, but winning after winning is the hardest thing in sport. Complacency is just around the corner; it's an excuse to do all the stuff you've denied yourself when you're on the way up.'

What Beswick found at Old Trafford was unique and boiled down to those 'social architects' described by Steve Peters – personalities whose influence, whether it's exuded in a calm, authoritative fashion (Paul Scholes) or rather more forcefully (Roy Keane), is strong and galvanises the team.

'Take Roy. We got on very well. He told me that when he was lining up as captain in the tunnel ready to go out against Arsenal or whoever in front of 70,000 supporters, he got confidence from knowing he had prepared during the week, and this confidence then spread to the rest of the team. If I've prepared thoroughly, done everything I can, ticked all the boxes, I'm ready to go, he'd say. If we've not trained how I think we should and I've taken shortcuts, I have a big question mark over how I'll play.'

With that sort of attitude, even the notoriously wet Manchester weather could be harnessed to cultivate success rather than drown it. 'I used to go down to the dressing room, chat and try to get my finger on the pulse of the team: where they were mood-wise, energy-wise. One day Gary Neville went, "Come on boys, rainy Tuesday morning." I walked outside, and the practice was outstanding, especially given the weather. When they'd finished training, I asked Gary what he had meant. And he said that they believed if they could train like champions on a rainy Tuesday morning with all the excuses not to, Saturday would be easy. It was an insight into brilliant athletes and a brilliant collective.'

Beswick calls elite performers 'warriors'. They display untouchable self-belief and, importantly, an ability to be comfortable with feeling uncomfortable. They're also eager to learn, and it's why players like Keane credit Beswick with improving his mindset, while Gary Neville turned to him when his form suffered after a disastrous performance in United's second game of the 2000 World Club Championship when they lost 3–1 to Brazilian side Vasco da Gama. Beswick says foreign players like Ole Gunnar Solskjær and Jaap Stam were more willing to have a chat with him and it was their acceptance of him that encouraged the historically more reticent British players to follow suit. Beswick helped the players and many of them taught him, demonstrating to him what it takes to be the best.

▼ Bill Beswick (left) says elite performers display untouchable self-belief and are comfortable with feeling uncomfortable

'You're on show every week and criticism is harsh. But top performers like Giggs love the challenge and have no fear. They don't fear anything in the opposition or the environment. The only thing they might fear is themselves. It's very hard to be an extraordinary athlete and an ordinary person, and making that transition between the two can be very difficult.'

George Best and Paul Gascoigne are two obvious examples of players who used alcohol to replace the adrenaline rush stimulated by 60,000 fans chanting your name. It's a very real problem, says Beswick, particularly among older players on the verge of retirement, which is a population that he's increasingly working with. Jamie Carragher is one footballer Beswick worked with before the Bootle-born defender hung up his boots in 2013, having joined Liverpool's School of Excellence as a 10-year-old in 1988.

> Being in the final years of your career is like walking the plank. The problem is you don't know how long that plank is.
>
> **BILL BESWICK** SPORTS PSYCHOLOGIST

'I think he got it right. He left while he was still in the first team and he left in control. His was a great example of how to make that transition. One player told me that being in the final years of your career is like walking the plank. The problem is you don't know how long that plank is.'

Carragher is old school. However, Beswick, as Ferguson did so adeptly, has had to adapt his techniques to suit the millennial generation. Now, to instil the fighter

Home advantage?

WHAT IMPACT DOES THE HOME CROWD HAVE ON THE PERFORMANCE OF PLAYERS? AND CAN SCIENCE MEASURE WHETHER IT MAKES A REAL DIFFERENCE?

What links England, Chelsea and AZ Alkmaar? Yes, you guessed it – they're all known for having maximised home advantage. England famously won their only World Cup the only time they hosted the finals; Chelsea went unbeaten at home in the league for four years, or 86 matches, between 2004 and 2008; and AZ Alkmaar remained unbeaten at home in Europe for 32 games until losing 3–2 to Everton in 2007.

Players and managers often refer to the crowd as their 12th player, the enthusiasm and fervour of the home fans egging the team on to victory. But is that feeling supported by the stats? And if so, what are the mechanisms at play?

First things first, home advantage does help, although less than it used to. Way back in the 1895–1896 season home teams won 64 per cent of the 480 games played across the two professional tiers in English football. Fast-forward to the 2015–2016 season and that figure had plummeted to 41 per cent across the top four divisions, before bouncing back to 45 per cent the following season.

Despite the drop, teams were still 12 per cent more likely to win if they played at home during the 2016–2017 season across the four leagues than if they played away. This differential has been exhaustively researched with one of the key theories being that the home crowd influences refereeing decisions. Interestingly, a 2014 study by Dr Chris Goumas showed that crowd density has a greater impact than crowd size; in other words, a packed 10,000-capacity stadium has a greater impact on the referee than a half-filled 20,000-capacity stadium. He also showed that if the fans are relatively distant from the pitch, they have less influence on a referee than when they're right next to it. This is perhaps why West Ham United have struggled since moving from compact Upton Park to the sprawling London Stadium and its athletics track.

mentality, it's more about the visual than the verbal. 'The digital generation coming through needs a different form of illustration to have an impact. They're more YouTube than textbook, so I use a lot of film clips. I was down with Bristol City recently presenting to the players about what builds belief. So I played a video of this Norwegian kid jumping off a cliff from 3,000m with a pair of plastic wings in a plastic body suit. Question is: how does he get the belief to do that? Then the answers come. He trusts his coach, he trusts the team around him, he prepares... Players nowadays find that much more convincing than a PowerPoint slide.'

Of course, not even the most experienced psychologist or manager can guarantee success. Despite Beswick's work at Bristol City, just one win in the last eight games of the 2017–2018 season saw the West Country side plummet from a play-off position to 11th. When it came to it, Bristol City wilted in the psychological heat of the battle. But it's the physiological management of heat that we will consider next.

Another school of thought focuses on the players and their hormonal response to playing at home, particularly in terms of testosterone production. Testosterone can significantly dictate a player's physical and mental state. It's associated with, among many psychological traits, motivation and aggression.

'The players walk out to a cheering crowd and it raises their testosterone levels,' explains Jamie Pringle, former lead physiologist at the English Institute of Sport. 'The crowd's behaviour influences testosterone release, and then the physiological release impacts the player's behaviour. It's classic alpha-male stuff. If you put a footballer in a situation where they feel they can dominate, like when they have strong home support, they'll produce even more testosterone so feel even more in charge.'

That might manifest itself physically in sprinting more; psychologically they may take more risks. It's a phenomenon examined by Professor Nick Neave of Northumbria University, who showed that salivary testosterone levels in football players were significantly higher before a home game than an away game.

Neave also noted that, 'Perceived rivalry with the opposing team was important, too, as testosterone levels were higher before playing an "extreme" rival than a "moderate" rival,' which could explain Chelsea's 28-year unbeaten home record against Tottenham Hotspur before a Dele Alli double ended the hoodoo in April 2018.

'Ultimately, a lot of this is psycho-physiological responses to an environment,' adds Pringle. 'You could take an anthropological approach and say it's about a man protecting his tribe. But what you find is that the real top teams and players overcome a crowd booing them or perceived injustice. Physically and mentally they can override the natural environmental response.'

In terms of practical applications for away teams to boost testosterone, Pringle mentions the work of colleague Dr Liam Kilduff who showed that if you undertake high-strength work that taxes the entire kinetic chain, like a back squat, the increased testosterone response has favourable effects for three to five hours after. You'd just have to ensure that a system-priming session like that on the morning of a match didn't tip over into fatigue.

BEAT THE HEAT

It's 1994 and the Republic of Ireland are playing Italy in the World Cup. Liverpool's Ray Houghton has sent the Green Army into raptures with a left-foot volley from outside the 18-yard box that leaves goalkeeper Gianluca Pagliuca flapping humid air. In temperatures of over 35°C, Ireland dig deep to score a famous victory.

It was a great goal from Houghton and a resilient team performance, but, for many, the game was memorable for Jack Charlton, all sunburnt visage and oversized white baseball cap, attempting and failing to cool and hydrate his players. 'Jack Charlton's been going potty at the FIFA people because they wouldn't let him throw water out [to his players],' announced ITV's touchline reporter Gary Newbon, 'which is absolutely ridiculous in these conditions after 24 minutes.'

FIFA's policy up to then had been that drinks could only be consumed at the touchline. On a sweltering day in New York, Charlton complained, loudly, that those in the middle of the field didn't have the same opportunity as the wide players. After the game, then General Secretary Sepp Blatter delivered his usual pragmatic response to the incident: 'It's no coincidence that the only manager to complain was the one from Ireland. He does not have a problem with water, he has problems with officials at most stadiums and that is not FIFA's fault. If his players want water against [their next opponents] Mexico, they can have it. There is no problem and he should stop complaining.'

◀ Ireland's Jeff Hendrick cools himself down during a 2015 Euros qualifier match against Italy

Despite Blatter's swipe at the 1966 World Cup winner, FIFA ruled that water bags, but not bottles, could be thrown onto the field of play and consumed. Charlton responded that it proved his efforts on behalf of the team were correct. 'It's rubbish to claim that tending to the welfare of my players was not part of my job,' he said. 'They trust me to do what is right for them and I will continue to do that. We have been right in pursuing this matter and FIFA's change of heart proves that.'

FIFA took until May 2014 to revise their hydration policy so that players could enjoy a water break at the 30-minute mark of each half if the mercury reached 32°C on the Wet Bulb Globe Temperature Index, which takes into account factors such as time of day, cloud cover, wind, humidity and location. Even then it took a Brazilian court ruling to order FIFA to enforce the regulation in time for the following month's World Cup.

▲ Ireland manager Jack Charlton 'elegantly' cools Steve Staunton at the 1994 World Cup

Matches played in the heat are rare in the English Premier League with August and May pretty much the only times when supporters can turn up in T-shirt and shorts. It's a different story on pre-season tours where the likes of Manchester United, Tottenham Hotspur and Liverpool all migrate to countries baking under hot and humid conditions. It's similar with major international tournaments, though one of the more surprising decisions taken during Blatter's FIFA presidency – to deliver the 2022 World Cup to Qatar – took things to a different level. FIFA commissioned Michael J. Garcia, the chairman of the investigative branch of its Ethics Committee, to look into allegations of corruption within the organisation. The resulting 250-page Garcia Report found that FIFA's voters failed to take heed of the governing body's own inspection team, who warned that hosting the tournament in summer in Qatar, when temperatures can reach 50°C, would be 'high risk'. FIFA later announced that the tournament would be played from 21 November to 18 December 2022 – right in the middle of European football seasons – in a still physically demanding 30°C.

Dangers of heatstroke

'We're all aware how demanding a hot environment is. Playing football simply adds an extra physiological, psychological and perceptual load,' says Professor Mike Tipton of Portsmouth University, who has spent the last 35 years researching the areas of thermoregulation and environmental and occupational physiology. 'It's why we're currently working on heat-related strategies for the Tokyo Olympics in 2020.'

Scientists like Tipton recognise that playing in the cold isn't as demanding, purely because the nature of exercise heats up the body. It's simple biology. A player's core body temperature is around 37°C. During high-intensity exercise in a hot environment, this can easily rise above 38°C, leading to a performance decline and potential muscle cramps. Studies have shown, for instance, that a professional team's overall distance run drops by 7 per cent in the heat with high-intensity running down a significant 26 per cent. That has huge tactical implications, which we'll come to shortly.

Performance impairment is secondary to physical damage when the mercury rises to around 40°C; at this point a footballer could be crippled by heat exhaustion, leading to nausea and headaches. In May 2017, England and Houston Dash forward Rachel Daly collapsed during the final minute of a National Women's Soccer League match against the Seattle Reign as temperatures climbed into the late 30s. She was taken to hospital and treated for heat exhaustion. Thankfully, she was able to be discharged that night. 'Everyone knows I'm a player who will give everything, but these conditions aren't safe to play in at your maximum,' a shaken Daly said.

Once body temperature tips over 40°C symptoms can include confusion, nausea and visual problems. Immediate medical attention is needed to prevent brain damage, organ failure or even death. Thankfully, heat-related deaths in football are extremely rare. I can find no record of any in professional football, though one 12-year-old American boy tragically passed away in 2016, two weeks after suffering heatstroke in practice that resulted in severe brain damage. (It's a different story in American football where heavy kit and longer games are two key causes why, from 1995 to 2015, 61 gridiron players – 46 high school, 11 college, two professional and two from organised youth matches – died from heatstroke.)

▼ England's Rachel Daly (centre) collapsed through heat-stroke when playing for Houston Dash

When you start to overheat, the body tries to cool down in four ways: radiation, where the heat generated from the body is given off into the atmosphere in the form of thermal radiation; evaporation – the process of losing heat through the conversion of water to gas; convection, which is where you lose heat through the movement of air or water across the skin; and conduction, where you lose heat through physical contact with another object – a football shirt or shorts, for example.

For these mechanisms to work effectively, especially when exercising in high temperatures, it is essential to stay hydrated. That's where the team's support staff come in.

Monitoring hydration

Manchester City's impressive training facility in East Manchester includes a nutritional room for the sports scientists and physiotherapists to prepare recovery drinks, pre-training drinks and specific supplementation.

On the wall is a poster with the number '17' shouting loudly in enormous type. It refers to a study that showed 17 per cent of footballers are dehydrated. There is also a line chart showing the hydration status of City's first team players, assessed via a simple test to measure the specific gravity of the players' urine (i.e. the density of the urine in relation to water). The denser the urine, the less hydrated the player.

> Your thirst mechanism switches off before you're fully hydrated.
>
> **PROFESSOR MIKE TIPTON**
> PORTSMOUTH UNIVERSITY

A simple daily visual check of urine colour is also the norm – brown urine means you are severely dehydrated – especially at hot tournaments, while measuring unclothed bodyweight is common before and after training or a match to measure water loss is common, too. 'The best teams we've worked with all do that,' says Professor Tipton, 'plus they measure the support teams as well, the rationale being that if the support team aren't performing at 100 per cent, they can't fully support the players. It's something you really need to do in a hot environment.'

Tipton uses chilled drinks with a hint of flavour, as he finds players are more likely to drink these than just plain water. Regular sipping is better than downing a pint-and-a-half in one go, as research shows that if you drink too much too quickly, the majority of the liquid will just flush through. It seems obvious that if it's hot the players will drink more. Not so, says Tipton. 'The problem is that if you leave people to hydrate, they tend to do so only to around 75–80 per cent of what they need. Your thirst mechanism switches off before you're fully hydrated.'

Research supports Tipton's assertion that left to their own devices, players come up short. A 2005 paper in the *International Journal of Sports Medicine* by Dr Susan Shirreffs and her team saw them collect data from 26 professional footballers during one approximately 90-minute training session played in 32°C. During training, the

▲ Manchester City pay significant attention to hydrating their players, like keeper Ederson

players had free access to Gatorade and mineral water, but only consumed an average 932ml compared to average sweat loss of 2,193ml (or 1,492ml an hour). Collectively, either through habit or simply the impracticality of constantly sucking fluid from a water bottle between drills and training matches, dehydration levels remained at a 1.6 per cent loss of bodyweight.

And that's not great because when you sweat, you lose water from extracellular fluid like blood plasma, which makes up around 55 per cent of blood volume. In turn, blood volume (about 5l) drops. That reduces cardiac output – namely, the amount of blood pumped each minute. So if Dele Alli is pumping out 100ml of blood each beat and his heart rate is a relatively relaxed 100bpm, the Spurs midfielder's cardiac output is 10,000ml or 10l per minute.

Now let's say Alli's blood volume has plummeted because of increased sweat rates, meaning he's now only pumping out 40ml blood per beat. Even when his heart's working harder – say 150bpm – his cardiac output is only 6l per minute. As blood delivers oxygen to working muscles, this is a problem.

The International Olympic Committee's official statement on the matter, based on current research and expert input, says that if an athlete experiences 2 per cent or more bodyweight loss through dehydration, performance is impaired. A kilogram equates to around one litre. If a player sweats 3l over the course of a game and doesn't replace it (we're talking hypothetically here as clearly that would be madness!), it would drop their, say, 70kg starting weight to 67kg – over 4 per cent of their bodyweight. Physiologically that would result in a greater cardiovascular and metabolic strain. Movement, intense runs and decision making would all become more difficult.

So it's clearly important to avoid getting dehydrated. For a match in normal British conditions the advice is to drink around 6ml of fluid per kilogram of bodyweight every two to three hours, so roughly a 500ml bottle for an 80kg defender. And all topped off with some electrolytes…

Don't sweat the white stuff

Andy Blow is co-founder of Precision Hydration, a company that specialises in sweat. 'We've worked with Bournemouth, West Ham, Newcastle, Sunderland, Norwich…' says Blow. 'We've also done stuff with Chelsea, numerous MLS clubs…' You get the point – the credentials are there.

Why are professional football clubs turning to Blow? Because they believe his sweat test, bespoke hydration strategies and in-depth knowledge of electrolytes can help their players to perform better, especially in the heat. 'Sodium is the key mineral in extracellular fluid because, among many functions, it regulates fluid balance by channelling water from the bloodstream to muscles via the sodium–potassium channel,' explains Ironman athlete and sports scientist Blow. 'So to enter working cells, water has to piggyback on sodium or it'll simply be weed out.'

> Most diets include sufficient sodium for the casual exerciser, but not for professional sportspeople.
>
> **ANDY BLOW** PRECISION HYDRATION

Sodium is vital then as a conduit for water, particularly so in the heat, but if Western diets are overloaded with salt, surely there's no need for a top-up? 'We've sweat-tested athletes and concentrations vary from 200mg to 2,000mg sodium per litre of sweat, which is purely based on genetics,' says Blow. 'Most diets include sufficient sodium for the casual exerciser, but not for professional sportspeople.'

Tell-tale signs of heavy sodium loss are white marks around the mouth or on the footballer's top – both more visible in hot conditions. If you're sweating huge amounts of sodium, drinking water alone could struggle to hydrate you because it will not be able to piggyback into the cells. That's where Blow comes in.

He tests footballers in pre-season when the weather's hotter, training is often more intense and prolonged, and the players are potentially less well conditioned.

First he asks them to answer a questionnaire to find out what they know about hydration, their own sweat rate and cramping history – do they have a history of cramping in hot weather or when sweating a lot? It helps him paint an individual picture of each player.

Blow then attaches electrodes to the players' skin to artificially stimulate sweat glands. The droplets are collected and analysed. 'We sold one machine to the FA, who have one at St George's Park [England's training centre],' says Blow. 'They're useful because you can then analyse sodium concentration to see if it's low, normal or high.

'We'll then risk assess and make a suitable intervention. Will someone have more of a problem under heavy training load? Or a problem in the heat? We tend to give the player two strategies. One we call a normal strategy, which is for the normal conditions they train and play in. In the UK in Celsius terms that's roughly the early teens. If they're then playing in hotter conditions or the match has gone into extra time, we give them another strategy that requires an extra supplementation of electrolytes. That could be for when they're in somewhere like Miami as part of a pre-season tour, for example.'

The lowest sodium level Blow has ever measured is 200mg sodium per litre of sweat, with the highest being 2,000mg, though he says a club's normal range is from around 500mg to 1,700mg. As he explains, 'Your sodium level's predominantly dictated by genetics, though it is affected by heat acclimatisation and, to a degree, fitness level.'

One might expect those of African ethnicity, for instance, to experience lower levels of sodium loss through sweat, in order to stay more hydrated in the heat. Blow thinks there probably is an ethnicity difference, but suggests more research should be undertaken in this field. And, in fact, data available from American football shows little difference between the sodium levels of Caucasian and African-American players. 'It's complicated. Take Turkey or the Middle East. There's a high prevalence of the cystic fibrosis gene, which results in higher sodium losses. So that's nothing to do with being born and growing up in a hot, dry climate. Also, different people's sweat glands are better or worse at reabsorbing sodium and chloride ions after leaving the body.'

Blow and Precision Hydration aren't the only ones focusing on sweat content. Over to Ajax's head of nutrition, Peter Res. 'Every three months we'll take sweat measurements by placing a patch on the players' backs. It collects sodium,' he says. 'You can then extrapolate that information for matches and training to individualise electrolyte drinks. We have scales in the locker rooms for players to weigh themselves before and after the match, too, but not every player is that interested all the time…'

Res highlights that though it's more important to monitor sodium loss in the heat, some players should consume an above-normal level of sodium all year round, even during a northern European winter, simply because they naturally sweat a high amount of salt.

Cool clothing

SPORTSWEAR BRANDS LIKE NIKE AND ADIDAS SPEND MILLIONS MARKETING KIT THAT'S COOLING AS WELL AS COOL, BUT THERE'S SOME SUBSTANCE TO THEIR PERFORMANCE CLAIMS

30 June 1998, Stade Geoffrey-Guichard, Saint-Étienne. England are level with arch-rivals Argentina after 16 minutes when 18-year-old Michael Owen collects a chipped pass from David Beckham just inside the Argentina half and unleashes the after-burners, storming past defenders José Chamot and then Roberto Ayala like they were statues before firing beyond goalkeeper Carlos Roa to put England 2–1 up. They'd lose on penalties but the sight of Owen bursting on to the world scene lives long in the memory. However, committed material technologists might be distracted from this beautiful goal by the sight of Owen's incredibly baggy top. Yes, the three lions flapping on a sultry French evening may have fanned him somewhat as he closed in on goal, but the design completely failed to harness the phenomenon that helps explain why today's kits are so form-fitting: wicking.

▲ Michael Owen of England on the ball during the World Cup group G game against Tunisia in 1998

Wicking is the process whereby moisture is drawn away from the skin and, in this case, to the surface of the football top. Because water conducts heat more than 20 times quicker than air, if sweat is left to pool on the skin, your skin temperature rises, which ultimately leads to a rise in core temperature and fatigue. 'That's why one of the key challenges in keeping comfortable is to reduce the amount of water that sits on the surface or in the fabric. Evaporated cooling is a constant target,' says Dr Simon Hodder, an ergonomist at Loughborough University who has worked with a number of top sportspeople.

Brands give their wicking fabrics flashy names – Nike Vapor (as seen opposite) and Adidas Climacool – but essentially they're just polyester. That's because polyester absorbs very little water – around 0.4 per cent of its weight – compared to cotton football shirts of times gone by that would soak up around 7 per cent of their weight. Polyester has a much higher density of fibres, which allows it to swiftly draw sweat through to the outer surface of the shirt for evaporation. But the right fit is essential.

'If a garment claims to be wicking, it needs to be in contact with the body,' says Hodder. 'If it's too loose, it can't do the job the manufacturer says. Then again, if it's too tight, you'll change the optimum fit. If you hold a wicking top against the light and pull it tight, you'll see what I mean. At a certain stretch it all seems to be together. Stretch it too much and you see light through it. They base it on yarns per square cm. Once you pull it apart, you decrease that number substantially, which reduces the effectiveness of wicking, as well as making the garment uncomfortable and restrictive to wear.'

The form-fitting designs, manufacturers say, also add support in an effort to minimise injury risk. This is debatable and hard to vouch for through independent studies. More proven is that close-fitting garments make it much more difficult for opponents to grab hold of you as you run past them.

Acclimatising to perform

In the build-up to the 2014 World Cup, England headed to Spain and Portugal for a warm-weather training camp. As the Iberian sun beat down on the players weaving around the training-drill posts, there was something strange going on: the likes of Leighton Baines and Ross Barkley's dress code was more appropriate to a December training session at Everton's Finch Farm than a summer's day by the Atlantic.

'We did sweat testing today,' explained then manager Roy Hodgson. 'We have a T-shirt, a lightweight training tracksuit and on top of that the wet top. Three layers. The players went through today's session with sweat pads on. They're being analysed in terms of the extra heat that we tried to generate and how they mentally and physically responded. There are going to be uncomfortable moments and you've got to learn to be comfortable with that.'

They were also acclimatising. Temperatures on the Algarve during the late-May camp hovered around an unseasonably low 19°C – around 10°C cooler than the conditions England would face against Italy in Manaus three weeks later in their opening game. By layering up, England were artificially mimicking the heat strain they'd face in Brazil. After a brief return to Wembley where they beat Peru 3–0, they flew to their World Cup holding camp in Miami for friendlies against Ecuador and Honduras. Originally they'd pencilled in pre-tournament games in Washington and

▼ Luke Shaw and Ross Barkley train in Rio and the heat of the 2014 World Cup

▶ Before England flew to Brazil, they acclimatised in Miami

Boston, but they decided to switch to a climate that more closely resembled the heat and humidity of Manaus.

And with good reason. Acclimatisation is the most effective way to prepare a player more used to cooler climes for performing in the heat. According to an influential paper published in 2015 in the *British Journal of Sports and Medicine* entitled 'Consensus recommendations on training and competing in the heat', the aim of acclimatisation is 'to improve thermal comfort during sub-maximal, as well as maximal, aerobic exercise in warm–hot conditions'. In other words, to get players to perform to their optimum when both jogging and sprinting. This process involves managing numerous physiological changes, including increased sweating and skin bloodflow responses, expansion in plasma volume (leading to a greater ability to control blood pressure) and an improved ability to maintain a good fluid–electrolyte balance, so neither too dilute nor too dehydrated.

As the England team showed, there are two ways to achieve 'thermal comfort': ideally, you would train in the exact conditions you are preparing for, but if this is not practical you can artificially recreate these conditions. 'You can spend time in a heat chamber back in the UK,' says Professor Mike Tipton of Portsmouth University. England players did this at St George's before the Brazil World Cup. 'The problem with that is it can become logistically tricky because of player location to acclimatise the whole squad. Then again, you can just set up a "heat tent" [a closed-sided marquee containing heaters and exercise bikes] where the players can exercise, ideally for around 90–100 minutes at a time.'

If such facilities are not available, a popular compromise is to take a hot bath or sauna after a training session. 'To fully acclimatise,' explains Tipton, 'you need to increase your core and skin temperature. Now, training in a cooler climate still raises core but not skin temperature. In fact, we worked with the England rugby union team before they went on tour to South Africa and they were wearing sweat suits when exercising. The thing is, we don't know how compromised it is because no one's done a definitive study into some of these techniques, which is one of the things we're planning to do for Team GB before the Tokyo Olympics.'

▲ Peter Crouch's high surface area to volume is theoretically ideal in the heat

Train warm, live hot

That's why the pre-tournament training camp is vital; a place that mimics the conditions of competition. Studies show that acclimatisation takes between six and 10 days with the majority of adaptations – a decrease in heart rate and skin and core temperature, and an increase in sweat response – happening in the first week.

Take the 2014 World Cup. England arrived in Miami on 2 June. They played two friendlies before flying to Brazil on 7 June, seven days before their first match against Italy.

'There are pros and cons of holding a camp,' says Tipton. 'One of the problems is that you have to change the time you train. If you tried to do your standard training session at midday, it would be too hot and there would be a drop in performance because your training load would plummet. Hence, you'll switch to training when it's cooler – early in the morning and in the evening. The middle of the day will be when you'll start doing incremental heat exposure. Basically, you have to separate training from heat acclimatisation, whereas in a lab you're in your normal environment so you can just go into the heat when needed.'

This 'train warm, live hot' principle is the equivalent of the 'train low, live high' altitude-training concept where you train at sea level and spend the rest of your time at altitude. 'There are further issues surrounding air conditioning,' Tipton continues. 'If players are looking to acclimatise to the heat, should they turn off their air-conditioning units? We've concluded that players should use them overnight so they sleep well, but the rest of the time they should turn them off and immerse themselves in the environment.'

Assessing adaptation

So the aim of acclimatisation is to raise core and skin temperature to induce profuse sweating and increase skin blood flow, resulting in a catalogue of physical adaptations that are desirable when competing in the heat. Professional footballers, who train over 190 days a year and play in upwards of 60 matches, have been shown to acclimatise to the heat twice as quickly as a sedentary person. This is because their core temperature increases each time they exercise in their home country, so they're partly acclimatised already.

A simple monitoring of a player's resting heart rate is one method to assess how well they've adapted to the heat. Another is a less intense variation of the Yo-Yo test (see page 45). 'I've often undertaken the sub-maximal Yo-Yo test with teams I've worked with at hot competitions,' explains the test's creator, Jens Bangsbo. 'It's a sub-maximal undertaking so not exhausting, and players will go for certain times in relation to their maximal score.'

Bangsbo used the test with Juventus players when he was assistant coach there, and with the Danish national team, where he still works. 'It's a good way to see if players have adapted and acclimatised,' he says. 'We also measure their heart rate at sub-maximal exercise to see if it's decreased. (You see a higher heart rate in hot conditions when they first arrive and then after 10 days of acclimatisation they will have a lower heart rate.)

'And we did the same in hypoxic conditions when we [Denmark] were in South Africa for the World Cup [in 2010]. You must adapt to altitude like you do the heat. It's a useful test because you can do it as part of the warm-up.'

> The fitter you are, the more able you are to maintain a stable deep-body temperature.
>
> **PROFESSOR MIKE TIPTON**
> PORTSMOUTH UNIVERSITY

The speed and extent to which a player adapts is down to a combination of nature and nurture. The genetic component is out of the coaching staff's hands. For example, people with lighter builds for their height tend to deal better with the heat than their heavier-set counterparts, because they have a higher surface area to mass ratio – put simply, there's more skin to lose heat from. Following that well-established biological principle, you might expect England's squad for Qatar to consist of 23 Peter Crouches.

However, there are definite training factors that offset any genetic advantages or disadvantages. 'There's a very strong correlation between aerobic capacity and dealing with offloading heat,' says Portsmouth University's Professor Mike Tipton. 'The fitter you are, the more able you are to maintain a stable deep-body

APPLY THE SCIENCE

If you can't stand the heat ... get out of the tattoo parlour!

Tattoos are omnipresent at all levels of football, from Barcelona to Barkingside, but, according to recent research, they could be impeding your performance, especially in the heat.

Raheem Sterling flew to the Russia World Cup on the back of a Premier League-winning season in which he scored 23 goals for Manchester City. But that counted for little when he was pictured at England's pre-tournament training camp sporting a tattoo of an M16 assault rifle on his right leg. To Sterling, whose father was shot dead when he was two years old, the tattoo symbolised a pledge never to touch a gun and to do his shooting with his right foot instead. To some members of the public and anti-gun campaigners, the image was 'totally unacceptable' and 'sickening'. To a physiologist, it might represent an impediment to Sterling's sweat glands...

◀ Lionel Messi is one of the 50 per cent of pro footballers who have a tattoo... ▶ ...it's one of many decorating the Barça and Argentinian legend

Lionel Messi, Sergio Ramos, Dele Alli – it's estimated that over 50 per cent of professional footballers now have tattoos, and many amateur players have followed suit. Responding to this global phenomenon, in 2017 Maurie Luetkemeier, Professor of Physiology and Health Science at Alma College in Michigan, USA, led a study into the impact of tattoos on a sportsperson's physiology.

The researchers applied small chemical patches to both the tattooed and untattooed skin of 10 fit, healthy young men who had a tattoo on at least one side of their body. These patches contained pilocarpine nitrate, a substance that stimulates sweating. The volunteers wore these patches for 20 minutes before Luetkemeier and her team removed them for analysis.

The findings revealed that the patches above the tattooed skin were much lighter; as it transpired, they'd absorbed half as much sweat as the patches on the untattooed side. The composition of the sweat was also different, perspiration from the tattooed skin containing nearly twice as much sodium as that from the untattooed side.

'That finding suggests that the underlying cause of the shift in sweat probably involves permanent changes within the skin after tattooing,' Professor Luetkemeier said. 'Perhaps some of the remaining dye blocks some of the sweat glands. But more probably, lingering inflammatory cells [from the initial tattoo itself] change the chemical environment within that area of the skin in ways that slow the response of the glands and affect how much sodium is incorporated from nearby cells into the sweat.'

Professor Luetkemeier concluded that it's 'unlikely' tattoos would impede perspiration enough to contribute to overheating or other problems even when playing football. Less circumspect was Dr Ingo Froböse of Cologne's German Sports University. 'I would forbid footballers from being tattooed,' Dr Froböse told the *Sun* in August 2017. 'Various studies have shown that players suffer a 3–5 per cent drop in performance after having a tattoo. The skin is the largest organ we have yet we poison it.'

Dr Froböse failed to elaborate on the studies that demonstrated such a performance drop, but he did note that 60–70 per cent of the ink from tattoos passes into the bloodstream. 'As a result, one's powers of recovery suffer and you are no longer as fresh as before.'

Further research is required to investigate the impact of tattoos on sweat rate, its composition and any potential detrimental effect on sporting performance. But one thing's clear: at this rate, within the next few years every footballer will sport a tattoo, meaning it'll be a level perspiration playing field come the 2022 Qatar World Cup!

temperature. We know that if someone's working at half their aerobic capacity, their core temperature is around 38°C.'

And that's important. Say England's Dele Alli had twice the aerobic capacity of France's Paul Pogba. Alli could do twice the work Pogba could do while maintaining a comfortable core temperature. That would mean he could outrun him, execute more repeated sprints and generally give himself more time on the ball.

That's a hyperbolic – and hypothetical – example, but it highlights how aerobic capacity is arguably more important in the heat than in cooler climes.

Heat-based selection

Professor Tipton feels that aerobic capacity and how a player adapts to the heat should have a far greater impact on team selection for matches in hot conditions than it currently does. He recommends, though concedes it's rarely done, that players undertake the heat-acclimatisation process 12 months before the competition to assess how they react, such is the variability in how players tolerate, acclimatise to and perform in the heat. Based on a heat-management standpoint – 'I'm not a coach,' he insists – the results should then impact selection and strategy.

'We base our definition of playing well on Manchester United playing Liverpool in the winter. That affects who we regard as good players and how we play the game. Typically, if you're playing in a 10 or 11°C environment in the UK, heat is not a limiting factor. So a player can run 11–12km in a game and a wing-back can sprint up and down the line all day. But players chosen for their energetic play won't necessarily be able to replicate that in the heat.

'It's the same with tactics. If your game's all about pressing, pressing, pressing, will you still be able to do that in the heat of Qatar? Probably not. Look at the Brazilians. There's a reason their play is more possession-based than focused on running hard and long.

> If your game's all about pressing, pressing, pressing, will you still be able to do that in the heat of Qatar? Probably not.
>
> **PROFESSOR MIKE TIPTON**
> PORTSMOUTH UNIVERSITY

'Ultimately, you've got to contextualise it. For every unit of energy you consume, 80 per cent is released as heat – roughly the same as for a petrol engine. That's fine in the cold but not in heat and humidity. You need a slower, passing, ball-control game.'

That's certainly reflected in a 2010 study led by Professor Magni Mohr of the University of the Faroe Islands, which analysed a range of football-specific metrics relating to 20 professional players before and after a game played in 31°C. Mohr noted that high-intensity running dropped by 57 per cent during the last 15 minutes of a match compared to the first 15 minutes, while repeated sprint and jump data dropped by 2.6 per cent and 8.2 per cent, respectively, as the game wore on. Which is hardly surprising as fluid loss during the match averaged two litres, while some players registered a muscle temperature of 41.5°C.

Pre-cooling strategies

Football clubs and equipment manufacturers have experimented with various ways to keep players' body temperature down in the heat. The highest-profile product in the elite sporting armoury is the ice vest. These cooling garments are well established in football; England players wore them at half-time in matches during the 2002 World Cup in Japan and South Korea. They began life as bespoke tools forged in the physiology labs of national sports institutes and soon spread around the globe. Adidas, for example, launched a new cooling vest for Spanish, Argentinian and German players to wear at the 2014 World Cup.

Adidas made great claims about the adiPower pre-cooling vest – that it reduced body temperatures and delayed the onset of heat-induced fatigue. 'It should be worn as players warm up for matches or during half-time intervals,' they announced. 'The garments are stored in a freezer to maintain a temperature of close to freezing for 15 to 20 minutes, allowing players a suitable cooling period whilst playing in Brazil.'

It's a commonly used tactic but one Professor Tipton questions. 'There's mixed evidence [for ice vests], certainly in a game of football that lasts over 90 minutes. If you pre-cool someone too much, muscle function deteriorates, which isn't great, especially at the start of a match. It can also bring about something called cold-induced diaresis. Everyone knows that if you stand in the cold, pretty soon after you need a wee. That's because you're shifting fluid from the periphery – because you're cold – to the core. The body senses that as an overload and you produce urine. So you're actually causing dehydration.'

Instead, Tipton recommends that teams transform their changing rooms into cooling stations featuring fans, cooling drinks and a hand-cooling facility. 'The ideal is that the players can dip their hands into cold water, but giving them a couple of ice-cold water bottles to hold is an acceptable compromise.'

Our hands contain a huge number of thermoreceptors and have a particularly high surface area to mass ratio, meaning they lose heat much more quickly than other parts of the body. Cool your hands and your brain receives a large amount of sensory information that makes the rest of you feel cooler, too.

A team led by Yang Zhang of Alabama University investigated this cooling phenomenon in 2014. They had seven footballers run for 45 minutes on a treadmill in a heat chamber set at 30.5°C. During a simulated half-time break, the subjects then recovered with three cooling options: passive cooling, which was simply stepping outside the chamber; arm and hand cooling outside the chamber; or neck cooling (again, outside the chamber).

They then performed a 6 x 15m sprint test and a Yo-Yo test outside the chamber. The players who had undergone the active cooling procedures scored more highly in both tests, which suggests that these methods improved comfort and sweat response and so delayed heat-induced drop-off in performance.

The disproportionate effect of hand-cooling was the principle that in 2013 led

Hitting the heights

ALTITUDE TRAINING IS COMMON AT MANY CLUBS, BOTH FOR IMPROVED PERFORMANCE AND ACCELERATED RECOVERY FROM INJURY

Manchester City's Etihad training ground and England's St George's Park both feature altitude chambers. But why? 'Regular exposure to altitude can lift performance by around 2 per cent,' says Sam Rees, Performance Specialist at the Altitude Centre in London, whose clients include Fulham, Liverpool and Aston Villa. The premise behind altitude training is relatively simple: by exposing a footballer to an environment that's low in oxygen, the body will adapt by becoming more efficient at transporting and using oxygen.

Air contains 21 per cent oxygen. But as you go higher, the air becomes less compressed, thinner and harder to breathe. So while footballers at sea level inhale 21 per cent oxygen, when they play at around 2,300m above sea level, the effective oxygen content of air decreases to below 16 per cent.

This can have benefits – as we'll explain shortly – but it can also cause problems, as Brazilian club Internacional found in March 2015 when they played Bolivian club The Strongest in their Hernando Siles stadium, situated 3,637m above sea level. Journalist Ewan MacKenna reported the match for the *New York Times*:

▼ Manchester City's Etihad training ground features an altitude chamber

Internacional had planned to spend only 12 hours in La Paz to minimize the effects of the city's altitude, but even that was plenty for the environment to take its claustrophobic toll. Just 36 minutes into the game, Internacional's star midfielder, Anderson – a signing from Manchester United on which the club has mortgaged a chunk of its future – was substituted so that he could stumble to the bench and put on an oxygen mask. Slowly, at least a half dozen of his teammates drained as well, and eventually Internacional lost, 3–1, to its Bolivian host.

Altitude had been enough of a factor for FIFA to briefly ban most matches there in 2007, though the ban was later lifted.

What Internacional failed to appreciate, and what Rees stressed, was that altitude exposure needs to be 'regular' for the body to adapt to the physical strain that, ultimately, results in better performance. Studies have shown that regular training at altitude triggers a number of physical adaptations that ultimately increase VO_2 max (see page 52) by anything from 3 to 8 per cent; decreases heart rate, both at rest and during exercise; increases levels of myoglobin, the muscle protein; reduces lactic acid build-up; and stimulates the kidneys to produce EPO (the hormone erythropoietin, which promotes the formation of red blood cells).

Altitude training is a common tool in endurance sports like road cycling and marathon running where the aerobic adaptations it promotes – particularly maximising oxygen delivery to the muscle – are hugely beneficial. However, football is a heavily anaerobic sport so, unless you're preparing for a match in La Paz, altitude training would seem to be unnecessary.

'That's a common misconception,' says Rees. 'Repeated sprint ability is probably the best marker of "fitness" for a footballer. Our research shows that sprint training in hypoxia [reduced oxygen state] accelerates recovery of heart rate between intense sets when playing on normoxia [normal conditions].'

Rees says that teams will often train in hypoxia (around 13 per cent oxygen) during pre-season when a lack of competitive matches means there's more time to train and recover. Generally, they would undertake three 20-minute sessions a week over a four-week period. In the case of, say, Manchester United and Tottenham Hotspur, who have just had a second altitude chamber installed at their Lodge training ground, the sprints are on static bikes as treadmills would be too dangerous.

The Altitude Centre also worked with the England football team prior to the 2010 World Cup, supplying tents that the players could sleep in to simulate altitude at night-time. They then trained at sea level. This is known as 'train low, sleep high'.

'Some Premier League players still use them for sleeping in,' explains Rees, 'though it tends to be to maintain fitness when they're injured, so that they can compensate to some degree for the reduced training load.'

The Altitude Centre also supplies clubs with portable units that players can use at home while exercising. Unlike the altitude chambers at clubs – where the entire room is kept at a reduced oxygen level – these consist of a mask connected by tube to a mini-generator that replicates high-altitude conditions. 'Clubs such as Norwich, West Bromwich Albion and Nottingham Forest use the portable units,' says Rees. 'They are popular with clubs that don't necessarily have the funds for a chamber but want to enjoy the benefits of hypoxic training. Clubs that use them, and footballers at home, will generally use them on the bike rather than the treadmill.'

two biologists at Stanford University, Professor H. Craig Heller and Senior Research Scientist Dennis Grahn, to launch the CoreControl Pro – cooling gloves that feel like having your hands thrust into a toilet cistern. However, they are more effective than that, because just putting your hands in cold water causes your blood vessels to contract, restricting blood flow and slowing the cooling effect. In contrast, when you use the CoreControl Pro your hands go into a vacuum, which keeps blood vessels open, meaning that the cooled blood travels more freely around the rest of your body. While its looks and bulk are clinical, the effects were promising, so much so that Germany used the system en route to World Cup victory in 2014.

Cooling from the inside

'We also have a heat protocol,' says Peter Res, head of nutrition at Ajax. 'Part of that involves cold towels before the game and at half-time, increasing fluid intake and consuming ice slushies. We used to have a machine that we'd install in the changing room, but now we simply bring them in a cooler box. It's more practical.'

That's right, in conditions over 26°C Klaas-Jan Huntelaar, Lasse Schöne, Dusan Tadic and co. will listen to manager Erik ten Hag's team talk while consuming a slushie drink. Studies have shown that consuming slushies can lower core temperature by up to 1°C through a process known as enthalpy, where the converting of ice particles into liquid removes heat from the body. This is why Res maintains that a slushie is better than cold liquid alone. 'The ideal slushie composition is around 60 per cent ice and 40 per cent water,' he says. 'You can also add some sugar for an energy hit.'

The concept arose in the build-up to the 2008 Beijing Olympics when a team at the Australian Institute of Sport recruited 12 top-level cyclists and began trialling new methods to reduce core temperature. After much experimentation, they discovered that athletes drinking 700–1,000ml of an ice slushie made by Gatorade realised a core temperature drop of 0.5°C. Even after a 30-minute warm-up, core temperature remained 0.5°C lower. The researchers also wrapped cold towels around the riders' legs and body but, despite showing positive results in their trial, this was thought to counteract the purpose of the warm-up because it cooled the muscles the athletes were trying to warm up.

Not everyone is convinced that slushies work. Some players find that swallowing solid ice to stimulate that enthalpy effect gives them stomach ache. 'I think ice slushies work best in short, really intense events, like a 10km time trial in cycling, where the core temperature rises extremely quickly,' says Professor Tipton. 'That doesn't really happen in football.'

The future is accuracy

Pre-cooling and acclimatisation strategies will be used by every national team with aspirations to do well in Qatar. However, at the moment it's hard to figure out which strategies will be most effective, given scientists have so little time to work with their national players to produce firm data.

Enter BodyCap. This is a French company specialising in miniaturised body monitoring devices. They recently worked with FC Nantes during two French league games and training to determine optimum protocols for their warm-up.

BodyCap's e-Celsius performance pill contains a temperature sensor, radio frequency antennae, four batteries and a processor, all wrapped up in a biomedical PVC shell. Simply swallow and then wait. Once it entered the gastrointestinal tract of the Nantes players, it began sending data (wirelessly, of course!) every 30 seconds to the software on a laptop.

'Because there's such a close correlation between temperature and performance, the Nantes team could use information from the BodyCap pill to devise individual warm-up and recovery programmes for each player,' says co-founder Sebastian Moussay. 'The pill is also useful in extreme environments like the heat, where the support team can make specific clothing interventions or plan different hydration strategies.'

It's no great leap to see how the BodyCap could be used before, during and after a warm-weather training camp to measure adaptation. And for Moussay, an ingestible pill is just the beginning for miniaturised sensors. 'Depending on the parameter to be monitored, the right device could be wearable, ingestible or, in the future, implantable.' And after the pill has passed through the body? BodyCap say: 'It's not meant for reuse.' Very wise.

Now that the Russia World Cup has passed, all eyes are set on Qatar 2022 where, despite the tournament's move to November/December, temperatures will still tip over 30°C. For players based in northern Europe, who will be playing for their clubs up to that point, that's going to be a significant environmental shock. Acclimatisation training will be vital, as will bespoke hydration strategies, as well as educating potential squad members about what little things they can do in advance to prepare for performing in the heat.

It's vital, too, that this education isn't reserved for the players who currently light up the Premier League, Serie A and La Liga. At the time of the 2014 World Cup, Marcus Rashford was learning the ropes in Manchester United's academy. Four years later, he was playing for England's senior team in Russia. In fact, with England's under-17 and under-20 teams both winning their respective World Cups in 2017, it's feasible that, as long as England qualify, their squad could be even younger than the one in 2018, which was the third youngest (average age 26) in the tournament. With the likes of Manchester City's Phil Foden and Tottenham's Kyle Walker-Peters being talked about as stars of the not-too-distant future, it begs the question: where are these precious talents coming from and how do clubs spot and nurture them?

NURTURING NATURE

At some point in the 2018-2019 season Tottenham Hotspur will return to White Hart Lane, albeit version 2.0, after spending more than a season at their temporary Wembley home. The new stadium will house 62,062 supporters, making it the largest club ground in London. Standout features of a development that's predicted to cost £750 million include: a 17,500-capacity single-tier southern stand, designed to generate sound akin to Borussia Dortmund's famed Südtribüne; a retractable pitch that will seamlessly switch between football and American football; a Tunnel Club where for 'just' £19,000 per season, you can stand behind a one-way-glass wall to watch Eriksen, Alli and co. stride on to the pitch; and, most importantly, a cheese and wine restaurant. 'We are building step by step for our future,' boss Mauricio Pochettino pronounced in May 2018. 'We are preparing to improve, and we believe we can improve and be stronger.'

The innovative plans and sizeable budget reflect the club's ambitions to reach the top of the European pyramid. But it could be that a mere £30 million outlay they made several years earlier will prove to be even more important. That's what chairman Daniel Levy spent on the club's new training ground, which opened in 2012. 'I've had the chance to play at Real Madrid, Arsenal and Manchester City and, for me, this is one of the best training grounds in Europe,' former Spurs striker Emmanuel Adebayor purred at the time.

◀ Harry Winks is one of Tottenham's many academy players to graduate to the England national team

Player development

It's Tuesday 3 October 2017 and, after much back-and-forth, I've finally been granted access to interview Aaron Harris, Tottenham's head of academy sports medicine and sports science. The Australian has been at Spurs for over 10 years and is regarded as an authority on player development.

What surprises lie ahead, I wonder, while queueing on the M25, a necessary evil as the training ground nestles in the shadow of England's busiest motorway. This is a club that has developed seasoned internationals like Ledley King, Sol Campbell and Glenn Hoddle. More recently, Danny Rose and Harry Kane have made the transition seamlessly from academy to senior level. In fact, on the morning I roll up to the 77-acre site, it's announced that midfielder Harry Winks has been selected for England's final two 2018 World Cup qualifiers. They turn out to be drab affairs, but 21-year-old Winks shone in the second game – a 1–0 win over Lithuania. He has been at Spurs since he was nine.

▲ Harry Kane and Danny Rose are two more England players who made it through the Spurs system

Clearly they're doing something right. But what's the secret? I've been granted access to Harris, but not the training centre – that's off-limits. In the covert world of elite football, it's not a surprise; what is a surprise emanates from my own stealth tour from the car park to reception, as beyond the 15 training pitches I spy an 18-hole putting course, with the greens up to Open standard. Unusual.

Behind the flags is a 4G pitch covered by an ETFE (Ethylene Tetra Fluoro Ethylene) foil roof, a material similar to that used for the biomes at the Eden Project in Cornwall. Through the glass windows are badminton nets, boards with 1, 2, 3 and 4 marked on squares on the wall, and bats and rackets of all sorts. 'Those are for multisport,' explains Harris on welcoming. 'We'll talk about that later.'

Intriguing. So let's start with the basics – what's Spurs' secret? 'What I'd say we've done well with the younger players over the years is apply an individualised programme,' answers Harris. 'That approach has helped us develop a significant number of academy players through to the senior squads.'

The EPPP: the good and the bad

The EPPP is the Elite Player Performance Plan, a long-term strategy rolled out by the Premier League back in 2012 to develop a greater number and greater quality of home-grown players. Integral to the new structure was categorisation of academies, from one down to four. All academies work across three phases: Foundation (under-9s to under-11s), Youth Development (under-12s to under-16s) and Professional Development (under-17s through to under-21s).

The 'key performance output' of category one academies, the highest grading, is to 'demonstrate regular graduation of players into the Premier League and the wider professional game'. To achieve this, the Premier League require clubs to provide a significant level of support including sports science, and across the three age phases 'typical total coaching access' is up to 8,500 hours per player. Category four academies must 'demonstrate the ability to graduate players into the professional game' and only have to cover the Professional Development phase. A category one academy costs around £1.5 million per year to run, while a category four academy costs around £100,000.

When it was brought in, the EPPP attracted criticism from many lower-league clubs who felt that the system was geared in favour of the big clubs. In the face of the increased demands – and costs – of maintaining academy status, some small clubs, such as Wycombe and Yeovil, decided to fold their academies soon after the introduction of the EPPP (though both later relaunched them).

Part of the problem was the fixed transfer fee system that replaced the system by which academy players' transfer fees were set by an independent tribunal. For example, when Liverpool winger Sheyi Ojo moved to Merseyside from MK Dons as a 14-year-old in 2011 – a year before the EPPP came in – the tribunal arrived at an undisclosed fee believed to be in the region of £2 million. Under the new system, MK Dons might have earned less than £100,000.

As a guide, the fixed fee for a category one academy player above the age of 12 is £40,000 for every year they've spent at the academy up to the age of 16. For a category two academy player that figure drops to £25,000 per year; and £12,500 for a category three academy player. For players aged nine

Category one academies

Independent assessors grade the academies and at present 24 clubs have category one status.

PREMIER LEAGUE	CHAMPIONSHIP	LEAGUE ONE
Arsenal	Aston Villa	Sunderland
Brighton	Blackburn	
Chelsea	Derby	
Everton	Middlesbrough	
Fulham	Norwich	
Leicester	Reading	
Liverpool	Stoke	
Manchester City	Swansea	
Manchester United	West Bromwich Albion	
Newcastle		
Southampton		
Spurs		
West Ham		
Wolves		

◀ A then 14-year-old Sheyi Ojo moved to Liverpool for £2 million in 2011

to 11, the annual fixed fee is £3,000 irrespective of the category of the academy.

Clubs do sometimes break the fixed-fee structure: Manchester City recently paid £175,000 to sign 13-year-old Finley Burns from category three Southend United, believed to be a record for a player of his age. The young defender had been attracting interest from a host of Premier League clubs, but his family were reportedly swayed by City's impressive academy set-up. City had to pay this relatively high fee to fend off the competition; under EPPP rules, they could have paid as little as £34,000 (two seasons at £12,500, from 12 to 13, plus three seasons at £3,000).

Increased coaching time

The Premier League argued that the EPPP was all about giving players greater exposure to quality coaching. Before the changes, FA rules permitted clubs to have only three hours of contact time a week with nine- to 11-year-olds – the age period labelled the 'golden years of learning' by Dennis Bergkamp, who until recently was coaching in the academy of his first club, Ajax. In the 12 to 16 age group, contact time was limited to five hours a week.

On average, a young player at an elite club in Spain, France or the Netherlands enjoys 5,000–6,000 hours contact time from the ages of nine to 21. Until the EPPP was brought in, an English youngster benefited from only 3,760 hours.

'Now, for category one clubs, it's nearer that golden 10,000-hour figure,' says Mo Gimpel, director of performance science at Southampton, a club long regarded as having one of the finest youth set-ups in football, having produced the likes of Gareth Bale, Theo Walcott and Alex Oxlade-Chamberlain. Gimpel is referring to the 10,000-hour rule proposed by writer Malcolm Gladwell, according to which you need to put in 10,000 hours of practice (equivalent to around 20 hours a week for 10 years) to achieve success in your chosen field. As mentioned earlier, category one academies are required to provide 8,500 hours of 'total coaching access' per player.

It's not all about hours, though, says Tottenham's Aaron Harris. First-team manager Mauricio Pochettino is renowned for his intense pressing style of play. It's a template the academy also follows. 'Our younger players train with a lot of intensity and there's a big emphasis on quality rather than us saying we need you out there for three hours a day,' explains Harris. 'In fact, we disagreed with some aspects of the EPPP when it was rolled out, as we felt that 12–16 hours of training each week, on top of match play, was too much for our new scholars, who were just 16 years old.

'We debated it with them. If you look at some of the top academies in Europe, like Barcelona, Dinamo Zagreb ... they aren't doing those sorts of hours. Our players might do the same amount of physical work in seven to eight hours on the training pitch as another club who are training for 12–14 hours.

'We weren't going to compromise the quality of our work and potentially ruin our players with chronic injuries. There were physios at other academies who were telling me they'd had five or six players out with stress fractures because they'd increased their training load so much. Once you get chronic problems, it can be tricky for the players in the future.'

Our younger players train with a lot of intensity and there's a big emphasis on quality rather than us saying we need you out there for three hours a day.

AARON HARRIS TOTTENHAM

Growth and maturation

Managing training load is even more important during teenage years than in adulthood and that's down to physiological, psychological and emotional issues around growth and maturation. Dr Sean Cumming is a senior lecturer in the Department of Health at Bath University. He specialises in paediatric exercise science with his research focused on growth and maturation. He started working with the Premier League around three years ago after the collective heads of academies identified the difficulty of comparing potential recruits of the same age when they are often at such different stages of physical development.

'Age of maturity is an interesting question as it's largely down to genetics with around 20 per cent down to environment,' Cumming says. 'Humans are unique in having such a wide variation in pubertal maturation. Some hit puberty early; some relatively late. That means you can get quite a wide range of biological ages [age in terms of physical development] within a cohort of players of the same chronological age.'

And this is where the problems begin with talent identification. In groundbreaking research by Amanda Johnson, who was Manchester United's lead academy physiotherapist between 2000 and 2010 and is now head physiotherapist at the Aspire Health Centre in Qatar, she took X-rays of boys from the under-9 age group and found that their biological ages ranged from six to 12. Up until 16, there was a minimum five-year spread of biological ages within chronological age groups.

'This has big implications for academies, where players are judged against their chronological cohort,' says Cumming. 'Early maturers hit puberty earlier than late maturers. That means greater muscle mass, which translates as greater speed and power. That gives early maturers a huge advantage over late maturers of the same age.'

In 2015, the German career website Karrierebibel examined the squads of a number of men's German national teams, from under-15 up to Joachim Löw's senior side. You might have expected there to be a positive skew to players born between July and September, because birth rates in Germany peak during the summer. And you might have expected January and February, on the other hand, to be under-represented, because these have the lowest birth rates.

But that wasn't the case. It was found that players born during the first part of the year were in the majority compared to just 12 players celebrating birthdays in November or December. 'It's called the relative age effect and is the same in the Premier League and all leagues,' says Tony Strudwick, former head of performance at Manchester United. 'You tend to see an over-representation in academies of players born within the first three months [of the age group].'

While in the majority of European countries the age group into which a player is placed is based on the calendar year (1 January to 31 December), in England and the Premier League the bandings follow the school year (1 September to 31 August). It means a child born in September will have a physical advantage over one born the following August because he or she will have had nearly a whole extra year to grow and develop. Yet the two will be in the same football team.

In the Premier League academies this results in player recruitment being heavily skewed towards those born in the September–November quarter, which accounts for some 45 per cent of the intake, whereas those born in the June–August quarter make up just 10 per cent of recruits. There's a similar trend with regard to senior Premier League players, with 'summer-born' players least represented in teams, but the imbalance isn't as pronounced as at academy level.

The relative age effect is marked, but not as much as the difference between early and late maturers. 'The data shows that the emergence of bias towards early maturers starts from around 11,' says Cumming. 'That's important because you can't just go analysing player's birthdays and grouping them. But either way, by the time kids reach the under-17s, some 80 per cent of them are defined as early maturers. Late maturers account for more like 4 per cent. In fact, data from the Swiss Football

> By the time kids reach the under-17s, some 80 per cent of them are defined as early maturers. Late maturers account for more like 4 per cent.

DR SEAN CUMMING BATH UNIVERSITY, THE DEPARTMENT OF HEALTH

Association shows that late-maturing boys are pretty much screened out of the system by this time.

'It's a problem because if you were to look at early- and late-maturing boys, eventually late maturers will catch up and often be better as adults,' Cumming explains. 'But the problem is that clubs are having to make decisions when kids are 14, 15, 16... As a coach it's difficult to pin down whether a youngster is succeeding just because they're an early maturer or whether it's because they have more talent and potential.'

Predicting height

The EPPP and Premier League clubs have taken steps to rectify the imbalance, starting with identifying where a youngster lies on the maturation spectrum. 'There are loads of ways to predict adult height but we use the Khamis-Roche Method,' says Southampton's lead under-9s to under-16s strength and conditioning coach, Sam Scott. The child's predicted height is based on a formula that takes account of his mother and father's height, and the child's current age, height and weight. 'We then get a percentage of where they are now compared to their predicted height. So if the data says that a 6ft player is at 95 per cent of his physical development, that means he is predicted to end up at 6ft 4in. The closer you are to your predicted height, the more physically mature you're perceived to be.'

Peak height velocity – the period during which a young player's height is increasing at its fastest rate – is another important maturation predictor used by many clubs, including Spurs. Based on a child's age, gender, weight, height and sitting height (because leg length increases before torso length), it is possible to predict when the child's main adolescent growth spurt is going to take place. For girls this is most commonly between the ages of 12 and 14, whereas for boys it usually comes a little later – between 13 and 15 – but tends to last longer and be of a greater magnitude.

When former Manchester United lead academy physiotherapist Amanda Johnson undertook her PhD, she used X-rays to determine players' maturity. 'You're looking for things like width of growth plates,' she explains. 'The closer to maturity a boy gets, the smaller those growth plates become because they're absorbed into bone. The late maturers have larger plates.'

Using predictive methods like these, Johnson is confident that she helped to

Academy awards

WHICH CLUBS' ACADEMIES PRODUCE THE HIGHEST NUMBERS OF PROFESSIONAL FOOTBALLERS?

Consider this: 0.5 per cent of eight-year-olds at academies make it into professional football. That's just one in every 200 talented youngsters graduating to the full-time ranks. It's stats like that which had Mark Crane scouring the internet in search of the team that produced the most professionals. His son showed great promise and Crane wanted him to have the best shot at working his way through the ranks.

So he set about calculating the productivity of academies all around the country. Using the website footballsquads.co.uk he filtered the English-qualified players that had made at least one senior appearance, before finding out which academies these players had grown up in. He then collated and analysed the data.

As you can see opposite, Manchester United finished top of the table with 70 of their graduates playing professional football during the 2016–2017 season, 18 of those in the Premier League. Arsenal came in second at 66 with Chelsea and Tottenham joint third with 58.

Not surprisingly, almost all of the most productive academies have category one status, although League Two club Crewe Alexandra, who've carved a reputation for producing notable professional footballers like David Platt and Danny Murphy, came in ninth. League One side Charlton Athletic went two better in seventh. Both clubs run category two academies.

Despite having category one status, Swansea City's academy was second to bottom in the list, above only League Two Cheltenham Town, who had just one graduate enjoying professional playing time in 2016–2017.

save the careers of players who are now in the first-team squad. 'There were players who at 16 were still absolutely tiny,' she says. 'Well, I'd got all these growth measurements on them, saying they were small now but when they were 21 they'd be bigger. It gave the club confidence to keep them and they went on to play for England.'

Johnson doesn't name names. Her former colleague at United, Tony Strudwick, is more forthcoming. 'I remember we were patient with Jesse Lingard and played him down a year. It gave him time to mature.' Marcus Rashford is another one who was reportedly a late maturer. 'Marcus is taller than when I arrived 13 months ago,' his manager José Mourinho said in August 2017, when Rashford was two months away from his 20th birthday. 'He's three centimetres taller and has put some muscle on.'

X-ray is the most accurate method to stop late-maturing players slipping through the net. The problem is, clubs aren't allowed to X-ray academy players unless there's a good medical reason for it. Johnson was only allowed to do so because the X-rays were for her PhD, so were signed off by the ethics committee at Manchester University. 'For me it's a non-issue,' she says. 'The amount of radiation a child is exposed to in an X-ray is akin to watching television for a couple of hours.'

			Number of academy graduates playing at each level (2016–'17 season)					
RANK	CLUB	EPPP CATEGORY	PREMIER LEAGUE	CHAMPION-SHIP	LEAGUE ONE	LEAGUE TWO	NATIONAL LEAGUE	TOTAL
1	Manchester United	1	18	20	17	10	5	70
2	Arsenal	1	9	13	14	10	20	66
3=	Chelsea	1	6	9	14	16	13	58
3=	Tottenham	1	12	12	15	15	4	58
5	Everton	1	9	6	12	13	13	53
6	West Ham	1	8	12	13	8	10	51
7	Charlton	2	4	8	15	3	14	44
8	Liverpool	1	6	13	10	4	9	42
9	Crewe	2	2	5	7	18	7	39
10	Middlesbrough	1	3	9	12	7	6	37
84=	Yeovil	3	0	0	0	1	1	2
84=	Swansea	1	0	1	0	1	0	2
86	Cheltenham	3	0	1	0	0	0	1

Maturity-based strategy

Whichever method a club employs to ascertain maturation status, what happens next is key. Just as Manchester United did with Jesse Lingard, it's not uncommon for clubs to play a lad down in a younger age group if they recognise his potential is being masked by a lack of physical maturity. Or move him up if he's more mature than his peers. Having a clear strategy is essential.

Former Brighton academy physiotherapist Sam Blanchard recalls how the club used to plan young players' training around their physical development. 'We had a big whiteboard in our medical room, which we divided into age groups from under-9s through to under-23s. From under-13 to under-15, we subdivided into green, orange and red boxes. Players in the green box were either a year away from their growth spurt or a year past, so you knew they were anatomically in a good place to train. If they were in orange, they'd be around six months either side of that growth spurt and we'd monitor a little more closely. If they were in the red box that meant we'd probably seen them grow 3–4cm in a month so they were going through their growth spurt. We'd make this transparent to the parents and the coaches and we'd lay off the training intensity.'

Instead, Blanchard would advise greater proprioception (balance and body sense) work. 'During these spurts, you'd see a decrease in coordination and skill and increased movement variability, so they'd be clumsier. Really skilful players would suddenly have a heavy touch and wouldn't spot the passes they saw previously.

'So rather than involve them in drills that involve shorter, sharper passing or cutting in tight space, we'd use them more on the periphery and work on their coordination. Or passing longer balls but still working on their touch.

'The analogy I always gave the lads was that if you hold a pencil where the rubber is against your nose, you can direct where you want that pencil to point. If I got you to do that with a broom handle, all of a sudden that becomes heavy and moves in ways that you can't control. Their brains are still trying to control a limb that was, say, 80cm long but [almost] overnight is 2 or 3cm longer. Nerves and muscles that made that limb function haven't had time to adjust and lengthen.'

▼ Jesse Lingard is a successful example of a late maturer

Tottenham's Aaron Harris agrees. While physically more mature players might be doing more intense plyometric work, the Spurs lads in the middle of a growth spurt might undertake a gym session with balance balls. 'We teach them to run more efficiently as it's a skill like everything else,' he says. 'It might be improving mobility of the hips or increasing strength and timing of your extensor torque – maybe your glutes – so that when you hit the ground, you push through and get into hip extension quickly. That potentially gives you a longer stride within the timeframe. We also use video with the guys, freezing the footage and asking peers to observe. You talk about speed equalling number of strides and length of stride. You are then going to get to the ball 1m faster. That could be the difference between scoring a goal and not. You always have to tie in the gym work with football. That's why if they're on a foam square, I want them to pretend to kick a ball, too. I want them to look over their shoulder; I want them to have someone behind them that they have to hold off. So it becomes more football-specific and you have more buy-in.'

The rise of bio-banding

At Southampton's Staplewood training ground, a path runs from the under-8s and under-9s area at the far end, past the academy pitches and then on to the first-team training pitches. It's a symbolic and literal pathway from youth to senior level. Walking down to the youth pitches with Mo Gimpel, the club's head of performance science, Gimpel points out a small area of tarmac. 'This is where the MRI lorry comes and scans some of the youth players,' he says. 'We're undertaking a study with Oxford University to help understand groin and hip injuries.'

It's another feather in the progressive cap for Southampton, who also hosted the Premier League's first bio-banded tournament in 2015, which also featured academy players from Reading, Stoke City and Norwich City. Instead of teams being selected on chronological age, they were chosen on biological age to level out the maturational playing field. 'In a normal age-group competition, the difference [in biological age] between early and late maturers is around 16 per cent,' says Bath University's Dr Sean Cumming, one of the leading drivers of bio-banding in the country. This means, for instance, that the most mature player might be 92 per cent of predicted adult height and the least mature 76 per cent. 'We said we wanted to restrict that to 5 per cent. You still get big kids and small kids with constitutional differences but you don't get that David and Goliath effect.'

The tournament was for 11- to 14-year-olds, with some 12-year-olds playing with 11-year-olds or 13-year-olds, for example. 'The kids loved it but for a variety of reasons. The early maturers playing up a group relished the physical challenge. They actually had to pass the ball and use technique to get around kids their size. In addition, they had to play more as a team, release the ball quicker and think a hell of a lot faster. It was an optimal challenge for them. "I couldn't just run through people and use my physicality" was a frequent comment.'

That last observation echoes a study by head of performance at Paris Saint-Germain, Martin Buchheit, who analysed the GPS data of academy players and, not surprisingly, found that the bigger boys ran more, at a higher intensity and undertook more sprints. But football is also technical and tactical, which is why late maturers also benefit from bio-banding.

'The late maturers didn't find it physically as challenging but they appreciated that because it helped them use their technicality,' Cumming continues. 'One lad said normally he'd be bulldozed by some massive lad. Now he can hold people off

The early maturers playing up a group relished the physical challenge. "I couldn't just run through people and use my physicality" was a frequent comment.

DR SEAN CUMMING BATH UNIVERSITY, THE DEPARTMENT OF HEALTH

and use his technical skills. "I was actually orchestrating the game," he said. "I was telling those younger players what to do."' Recent data from Exeter City also showed that in bio-banded games, there was twice as much passing and dribbling compared to age-group matches.

Cumming interviewed 48 kids and asked if they would like bio-banding integrated into their current programme. All but one said they would. Coaches are buying in, too.

Since the summer of 2015, a number of clubs have held bio-banding tournaments and the signs are positive. Cumming is clearly a fan, though he adds that they should complement not replace age-group competition 'because age-group matches are a great method for matching kids on the basis of experience and cognitive, motor and social skills, which generally follow age ... but hopefully bio-banding will keep more late developers in the system.'

There are critics, however. Some clubs refuse to bio-band because they worry that teenagers who are played down may be stigmatised socially or feel out of their depth if they are played up. 'It's why we took a four-corner approach to bio-banding,' recalls Sam Blanchard of his time at Brighton. 'As well as physical maturity, we looked at other factors like technical and tactical awareness, and social awareness. If you put an emotionally immature lad in an older changing room, the level of conversation and banter could be quite challenging. So are there safeguarding issues? And how are they viewed when they go back into their own age group? We had a psychologist gauge whether they were resilient enough to go up or down.'

This, according to Gimpel, is where role models come in. 'Alex Oxlade-Chamberlain is one of the best examples of a talented late developer. At under-16, we played him down a year and then offered him a scholarship at the club. Other parents were wondering how we could offer this boy a scholarship when he couldn't even get into the under-16s. But we knew in-house that the growth would come.

'When he hit his growth phase, within six months he'd gone from sitting on the bench for the under-18s to playing for the first team. He missed out the under-23s and is the only player I know who jumped that step. And then within six months he was at Arsenal. Now he's gone to Liverpool for nearly £40 million. To sell the idea of bio-banding, we call it the Alex Oxlade-Chamberlain pathway.'

Gimpel and Southampton also had to decide whether to put another late developer, Gareth Bale, on a scholarship. The diminutive Gimpel laughs that 'he was smaller than I am. Technically he ticked the boxes and had pace,

> The later maturers – if they stay in – can have more of a chance of getting through the academy system to senior level. They might be technically better. They're also mentally more resilient.
>
> **SAM SCOTT** LEAD UNDER-9S TO UNDER-16S STRENGTH AND CONDITIONING COACH, SOUTHAMPTON

but he wasn't tall or big and would get knocked off the ball easily. That was probably the first time in our academy we used maturation prediction. Using the established methodologies of the day, we predicted a final height of 6ft 1in and signed him. Whatever did happen to him…?'

The underdog theory

So three cheers for bio-banding? Not necessarily. Beyond the social-stigma issues, some argue that late developers build mental resilience and learn to overcome physical hurdles by playing against bigger opponents. 'It's called the underdog theory,' says Sam Scott, Southampton's lead under-9s to under-16s strength and conditioning coach. 'Basically, if you want to be a footballer and you're behind, it gives you a mental push. If you're already top of your game at junior level, it can be difficult to improve.

'With the underdog theory in mind, it showed a little in our stats that the later maturers – if they stay in – can have more of a chance of getting through the

▶ Southampton kept Alex Oxlade-Chamberlain (right) down a year. It worked

academy system to senior level. They might be technically better. They're also mentally more resilient.'

James Ward-Prowse is one recent example at Southampton, as is Harry Kane at Spurs. 'I would say Harry was a late developer, but he made up for that with his attitude,' says Aaron Harris of the forward who was let go by Arsenal before joining Spurs at 11. 'I can still remember him as a 15-year-old with a minor knee injury and he basically ruined himself in a rehab session. I did a lot of athletics training when I was younger and you're often coming off that track and you can't move. He had that mindset; he was mentally strong, a good lad and pushed himself to the limit from an early age.'

Is it also simply coincidence that the technically gifted Lionel Messi, Xavi and Diego Maradona were reportedly late developers? Messi actually had growth retardation, a condition where you don't produce enough growth hormone so you're small for your age. If left untreated, you'll be relatively undersized and not meet your growth potential.

▲ Gareth Bale was spotted and nurtured by the Bath arm of Southampton's academy

He was with Newell's Old Boys at the time. But at nearly US$1,000 a month, the treatment was too expensive for the Messi family and his club wouldn't pay. Having relatives in Catalonia, the family approached Barcelona for a trial and Barça were impressed enough to sign him and cover his treatment costs.

Growth and maturation expert Dr Sean Cumming accepts that some late maturers have better self-regulation skills such as weighing up their strengths and weaknesses and finding new ways to overcome their weaknesses. 'But, still, only 4 per cent of late maturers make it through the system – something's not right.'

Re-educating the academy decision makers won't happen overnight. Former Manchester United academy physiotherapist Amanda Johnson says that part of the problem stems from the scouts, many of whom are 'pure football men' or parents of academy players and, in her eyes, unqualified to make decisions based on complex issues like growth and maturation. That's seemingly not an issue in Catalonia.

'In academies like Barcelona, there's a big emphasis on the technical,' says Cumming. 'If you're focusing more on the technical than the physical, which is what we're trying to do with some of our work with the Premier League, you need to be able to separate them. I think they're very good at that in Spain. They produce a lot of skilful players.'

Testing times

ACADEMIES USE VARIOUS PHYSICAL TESTS TO ASSESS WHAT FITNESS ATTRIBUTES YOUNG FOOTBALLERS NEED TO WORK ON

Speed and power are increasingly important attributes in the modern game. 'It's why we test the youngsters' speed using a 20-metre sprint test,' Leicester City's strength and conditioning coach, Matt Wilmott, recently told *FourFourTwo* magazine. 'Every player is placed in a band ranging from poor to excellent [see below].'

Poor = 3.17 sec or more
Below average = 3.16–3.06 sec
Average = 3.05–2.96 sec
Good = 2.95–2.86 sec
Excellent = 2.85 sec or less

Coaches can then adjust players' training programmes to address any weaknesses. 'But we'd never use their physical tests to make an absolute decision on whether a player would get a professional contract. It's only to advise coaches from a physical point of view,' adds Wilmott.

As well as the 20m sprint tests, Leicester's academy players undertake the following (under-18s and under-23s, every six weeks; under-9s to under-16s every 12 weeks): vertical jumps to measure single- and double-leg power; a Yo-Yo test for endurance; and an arrowhead run to measure agility and explosive power.

An arrowhead run requires six cones laid out as shown in the diagram below. From the starting position, the player runs as fast as possible to the middle cone (A), before turning to sprint around one of the side cones (C) or (D), around the far cone (B) and back through the start/finish line. The test comprises four trails, two to the left and two to the right (as shown). Record the best time to complete the test for the left- and right-turning trails.

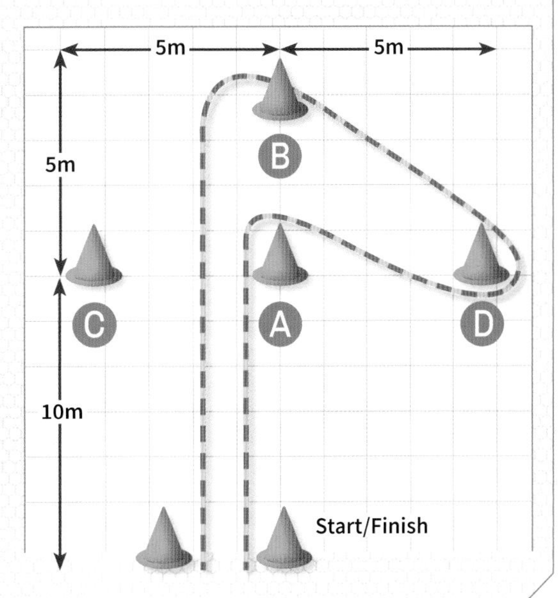

The Catalan talent factory

Nothing symbolises Spain's focus on the technical more than Barcelona's La Masia youth academy. It was first proposed by Johan Cruyff in 1979 to then club president Josep Núñez as he thought Barcelona should try to replicate Ajax's successful academy. Núñez agreed. The original academy building was an old farmhouse near Camp Nou actually called La Masia. Despite moving in 2011 to the club's new training centre, the academy is still often referred to as La Masia. The new academy base has living quarters for 83 academy players (out of 300 in the Barça system) plus two psychologists and numerous teachers and coaches.

While the façade is new, the soul and ethos remain the same. In Barcelona's impressive museum is a room dedicated to La Masia extolling the eight principles of the academy – the Barça way: 1 Individual and collective technical quality; 2 Rational occupation of the pitch; 3 Protecting the ball; 4 Mobility of players who don't have the ball; 5 Playing together as a team; 6 Overall understanding and interpretation of the game; 7 Individual and collective participation; 8 Individual qualities at the service of the team.

It's notable that the first principle focuses on the technical. Within that are sub-sections: constant contact with the ball from the start; work on coordination with and without the ball; technical content (what?) at different stages in training (when?); ideas and watchwords (how?) for improvement in all technical areas.

◀ So popular is futsal in Spain that Barcelona have their own professional team

▲ Barcelona's youth set-up is the envy of clubs around the world

This technical framework has guided the Catalans for years. Pep Guardiola arguably represents the success of La Masia more than anyone else, being one of the first major talents to emerge from the crumbling farmhouse. Guardiola the manager later built his team around graduates from the academy including Carles Puyol, Gerard Piqué, Andrés Iniesta, Xavi and Lionel Messi.

Under Guardiola, Barcelona won 14 out of the 19 major trophies available. His successor, Tito Vilanova, made history in November 2012 by fielding a first team made up entirely of La Masia graduates for an hour in a 4–0 win over Levante after Martín Montoya replaced Dani Alves in the first half.

Flourishing with futsal

Core to La Masia has been futsal. The five-a-side variant of football is so popular in Catalonia that Barcelona has its own professional team, FC Barcelona Futsal, who play within the shadows of the Camp Nou. The Football Association believe that futsal could hold the key to improving the conversion rate of academy players: currently just one in every 200 under-9 players will make it to senior level.

In the summer of 2017, the FA announced that £300,000 from the Football Foundation would be made available for schools, colleges and youth football leagues to create 200 futsal hubs. The aim is to introduce 12,000 more children to the fast-paced game that Neymar, Messi and Ronaldo credit with honing their skills as youngsters.

'Futsal in schools needs to happen,' declares Michael Skubala, the FA's futsal elite performance manager and head coach of the England futsal team. 'Until we start doing futsal properly, we probably aren't going to win a football World Cup like Spain or Brazil. These nations are doing it on a massive scale; all their kids are doing it. And that is giving them huge football returns later down the line.'

A recent study highlighting the differences between football and futsal using Barcelona youth players describes futsal as a 'laboratory' of intensive perceptual skill development. The study cites 'the higher game intensity, higher opponent pressure, an easier-to-handle ball and a lower number of players' as crucial to the game's benefits.

It's something many Premier League clubs are acutely aware of. 'We play futsal at Spurs,' says Aaron Harris. 'It's great technically and strategically as you're more accountable if someone gets past you. Your touch must be extremely good because there are so many people close to you and there's less margin for error. Physically it's a different game, too, as there's a greater bodywork component than on outdoor pitches.'

Harris elaborates that Spurs' academy players have played futsal all over Europe and won an under-17 tournament in Germany in 2016 against the likes of Bayern Munich. 'I expect the Premier League, who are pushing it, have seen the German model and tried to replicate it.'

No ordinary five-a-side

Futsal is the only FIFA- and UEFA-sanctioned five-a-side version of football. There are certain subtle but important differences from other five-a-side variants that force players to make quicker and better decisions in tighter areas:

- Heavier, smaller, less bouncy ball than standard
- Goals no bigger than 3 x 2m
- Fast surface (e.g. indoor sports hall)
- No bouncing the ball back in off the walls (there are touchlines like in 11-a-side football)
- Laws of the game must be followed strictly, which include a four-second rule on kick-ins.

International experience

International competitions like the futsal tournament Spurs won in 2016 are essential for player development. 'You can learn a lot more about these youngsters in four days away than you will at home in four to five weeks,' says Harris. 'You're seeing them for 24 hours a day. You're seeing their food decisions, their sleeping decisions, how they integrate socially in a different group. Can they hold a conversation with adults over dinner? That's a marker for when they're 19 or 20 and they have to go into another challenging environment, which might be a first-team changing room, and the potential stresses that might bring.

'I remember one year, Tom Carroll, who's now with Swansea, was told to mark Romelu Lukaku. Tom's a lot bigger now, but then he was eight-and-a-half stone, wringing wet and marking someone who was 12 or 13 stone – so 50 per cent bigger than him and faster. But we told Tom to use his football intelligence. It was about development, not winning. It's about making them feel uncomfortable. We use the acronym T-CUP – Think Calmly Under Pressure. Can you be comfortable with being uncomfortable, because it can't be comfortable being out in front of 80,000 people, 20,000 of whom might be booing you? Or facing an opposition crowd that hate you at the Emirates?' Clearly, this approach worked for Carroll as he has developed into a regular senior first-team player.

Amanda Johnson also emphasises the importance of international tournaments. 'When I was at United, we'd go out with the under-12s to a tournament every festive period. It was always in Spain and sponsored by Canal Plus – a sort of mini Champions League with the big Spanish clubs and PSG… Well, we got battered every year because we were one of the smallest teams.

'Madrid or Barcelona would always end up in the final. They'd often have these really big lads up front who just scored goals for fun because their heritage was African, and it's been shown that they mature earlier than Caucasians, who mature earlier than Asians. I noticed that in Manchester. But talking to their coaches, they'd say these boys won't make it. As for our lads, they learned from it, hopefully built resilience and moved on.'

Applying pressure is a careful balancing act. Too much too soon can lead to early professionalisation, which can interfere with young players' emotional and social development. This problem can be made worse if players spend too much time within the bubble of the academy. Some clubs even have a school on site, although not Spurs: 'We want our lads to identify as people not just footballers,' says Aaron Harris, head of academy sports medicine and sports science.

> We want our lads to identify as people not just footballers.
>
> **AARON HARRIS** TOTTENHAM HOTSPUR

In September 2017, the *Guardian*'s David Conn investigated depression and other mental health problems suffered by young academy players, particularly after they've been released. He quoted a study by the Professional Footballers' Association, which found that of the boys who make it into an elite scholarship programme at 16, five out of six are not playing professional football by the time they're 21.

It prompted Howard Wilkinson, former technical director at the FA, to call for a review of the system: 'Research shows that to have a successful national team, a country needs a group of around 50 players with high-level, including latter-round Champions League, experience, capable of playing in the team. England does not have that number of players who have been given the opportunity to go on and develop.'

Multisport excellence

At the annual Future of Football Medicine conference held in Barcelona in May 2017, Professor Roald Bahr, head of the Aspetar Sports Injury and Illness Prevention

Programme in Norway, told the attendees, 'I believe the United States has an advantage over many of the world's countries in that school sports programmes are limited to three months during the year, so I played football in the fall, athletics in winter and lacrosse in the spring.

'Many coaches like to say practise early if you want to be a great athlete in my sport. My question is: was Michael Jordan a good basketball player? He only played basketball three months a year. He also played baseball, football ... and this, I think, is an advantage for US athletes.' It was an observation supported by the recent NFL draft of college players, where all but three of the players chosen in the first round were multisport athletes.

There is a growing body of evidence that multisport helps not hinders sporting development, with the dangers of early specialisation highlighted by a number of studies, including a consensus statement published in the *British Journal of Sports Medicine* on reducing risk of burnout and injury in youth sport, which emphasised the benefits of diverse sports training during early to middle adolescence.

'I'm a huge believer in multisport,' says Tottenham's Aaron Harris. 'We have 30 minutes before every under-18 and under-23 training session and, ideally, within that we have 15 minutes of multisport. The sort of things that we do are badminton, basketball ... and a lot of rugby. But we tie it back in to specific skills the coaches want us to develop. So if they say today will be about improving passing with both feet, in rugby we'd have them passing the ball to both sides. Or have them running diagonally as football's not about running in a straight line.

'We also have an association with the NFL and a couple of times they've sent players to the training centre. NFL is seen as cool by the youngsters. They like the bling of it. So we might run through a Gridiron set play. We've been going out to Florida for the past four or five years, just prior to Christmas. So maybe in the build-up to that, we'll have less rugby, more American football.'

Harris says multisport provides different physical stimuli, which creates an all-round stronger athlete; and it also potentially cuts injury rates by avoiding heavy repetition of football-specific movements like kicking. 'Then there are transferable skills. Playing two-on-two basketball and working out your commitment to defend is like being a centre-half organising the back-line. Can you make sure your man doesn't run off you?'

Spurs also run six-week blocks from March to April where the under-9s to under-11s do multisport for an hour, twice a week. Handball, athletics ... but it's not enough, says Harris. 'I'd prefer 10–12 weeks, but this is a competitive environment. The club down the road will say to the parents we offer a 44-week programme, we also offer this much in travel costs, join us. Recruiting and retaining your best players is a big priority. You can't give everyone else a competitive advantage.'

And that's the crux of the situation: the theoretical ideal versus competitive and commercial pressures. And those are certainly fierce when you have clubs like

Tottenham and Arsenal fighting for the same youngsters. EPPP rules state that from nine to 11, academy recruits must live within an hour's drive of the training facility; that rises to 90 minutes from 12 to 16. It's why Southampton has a satellite centre in Bath – drive for five minutes one way from Southampton and you're in the English Channel, drive the other way and you're in the New Forest. Clubs also can't sign players from overseas until they're 16. One example is Paul Pogba, who Manchester United signed soon after he turned 16. 'Some might come with their family, others might stay with a host family,' says Amanda Johnson. 'Either way, it just highlights how competitive academies are.'

The globalisation of football has undoubtedly increased this competition. More than 60 per cent of Premier League players are foreign. That's a historic high and is seen as a major obstacle preventing home-grown youngsters graduating to the senior ranks. On the other hand, it's argued that training alongside some of the world's best players raises standards among young domestic players, making them more technically adept, fitter and stronger than ever before. It will certainly be interesting to see just how many players from England's victorious 2017 under-17 and under-20 World Cup squads become permanent fixtures at senior level for their clubs.

▼ Chelsea produce many pro players … but few make the first team

Heineken
BROWNHILL
8
Poplar Insulation
CITY

PITCHES, PADS, BOOTS AND BALLS

'Radford, now Tudor's gone down for Newcastle. Radford, again. Ooh, what a goal! WHAT A GOAL!' Twenty-six-year-old John Motson's succinct, euphoric commentary to describe Ronnie Radford's 30-yard piledriver that 'flew into the top of McFaul's net'. Radford's memorable goal set up fifth-division Hereford United for one of the biggest shocks in FA Cup history – a 2–1 win over first-division Newcastle United in 1972. It was the first time a non-league club had beaten a top-flight team in 23 years. It was also the fourth time the sides had attempted to stage the third-round replay after a particularly sodden January, ensuring a quagmire awaited. It was nothing new. In times gone by, these mud baths were the norm, so much so that four clubs – Luton Town, Oldham Athletic, Preston North End and QPR – opted to install artificial pitches in the 1980s. But with technology, investment and experience, the times they are a-changin'...

Pretty green

It's November 2017 and I'm at West Ham's Romford training ground. 'Half a million, half a million, three-quarters of a million,' head groundsman Dougie Robertson reels off in his deep Scottish accent as he points to each practice pitch in turn. 'That's another half a million over there, £800,000...' he continues.

On a crisp autumn Monday, as I survey the vivid green carpets all around, the investment certainly looks like it has paid off. 'I've been with West Ham for 20 years and the technology and pitch quality have evolved beyond recognition,' says Robertson. 'We have a mix of different pitches here and they're all top class ... and they'll remain that way all through the season.'

◂ Bristol City's Ashton Gate has benefitted from Desso grass

Radford's missile of Edgar Street was all the more impressive because he had

◀ Quagmires were the standard pitches of winter football in the UK

to overcome not only the elite opposition but essentially a ploughed field. Today's pitches are pristine, even at the training ground, and have played a major role in the swift evolution of the game with super-slick passing and unimpeded high-intensity bursts resulting in an ever-faster product.

The turning point for pitches in this country arguably happened in 1996. That's when Huddersfield Town's McAlpine Stadium (now called the Kirklees Stadium) became the first British ground to have Desso GrassMaster technology installed. This hybrid system, which weaves artificial fibres into real grass, proved so successful that it spread its blades from the McAlpine to the finest stadia in the land.

'The majority of stadia in the Premier League feature Desso GrassMaster,' explains Dan Prest, education and training manager for the Institute of Groundsmanship (IOG), 'but there are other hybrid products that teams can now choose, including SIS grass, Hero and Mixto.'

But it's Desso that is to professional football what Hoover is to vacuum cleaning, sucking up clients left, right and centre. Of the 20 clubs competing in the 2018–2019 Premier League, 16 have Desso pitches at their stadia and some, including Manchester United and Manchester City, also use the technology at their training grounds.

It's why Premier League matches are almost never postponed these days. 'And that's vital,' says Robertson. 'It's a Premier League ruling that you simply can't have a game called off. We kick off at 3 o'clock but we also must consider the Chinese audience, the American audience… If you miss kick-off, all the commercial rights for that period of the game have gone. And that's big money. Ultimately, we are groundspeople in a billion-pound industry.' In fact, on the rare occasions that games are postponed, it's generally down to icy conditions impeding routes into the stadia or making walkways slippery underfoot rather than the state of the pitch.

The importance of this hybrid turf can't be overstated. Take the case of the new Wembley. The FA turned to Desso in summer 2010 after having to relay the Wembley turf 11 times since the stadium reopened just three years earlier. Player after player had criticised the pitch; Michael Owen blamed the deteriorating turf for the ruptured hamstring he sustained in Manchester United's 2010 Carling Cup Final triumph over Aston Villa.

▼ Michael Owen blamed the turf when rupturing a hamstring during the 2010 Carling Cup triumph over Aston Villa

At England's HQ, three computer-controlled machines worked for up to 22 hours a day for nearly two weeks to insert 48,000km of polypropylene thread into the soil. Much effort, much reward, as Wembley finally has the world-class pitch befitting a national team. And it's all down to an industrial-sized sewing machine…

Grass-roots football

'This is what it looks like,' says Bristol City's head groundsman, Dan Sparks, who had Desso installed at Ashton Gate in 2014 and who is currently waving under my nose a huge needle attached to a roll of green fibre. 'The Desso is stitched 30cm down into the soil with 20mm above the surface. The fibres are around 20mm apart. When it's first sewn it's nice and tall, but then it weathers and curls up a bit. But that doesn't matter as it's what happens beneath the pitch that makes Desso so useful.'

While Desso might be responsible for the uniform greenness of once patchwork pitches, it's the stability those 20 million fibres provide that has transformed the way football's played in the modern era. Over time, the natural grass roots intertwine with the underground Desso fibres, which act as a form of horticultural scaffolding. The grass roots grow deeper and wider, forging a stronger and healthier root zone.

This gives traction beneath the surface, meaning slips and huge divots, often the cause of hamstring injuries and groin strains, are the exception rather than the norm. Although there are still plenty of these injuries (see chapter 4), it's no longer the pitches that are the main culprit. 'That's one of the key attributes of Desso,' says Sparks. 'If there's a slide tackle, for example, the turf holds together. If there were just sand underneath, the turf would fly off.'

Desso GrassMaster is now employed globally at 450 clubs across football, rugby and American football with the benefits felt particularly by those clubs in wetter climes. 'It's a sand-based system, meaning drainage is impressive,' says Sparks.

▶ The perfect Desso pitch at Arsenal's Emirates. The blue lines were to aid visibility in case of snow

▼ West Ham's Rush Green training ground has flourished under head groundsman Dougie Robertson

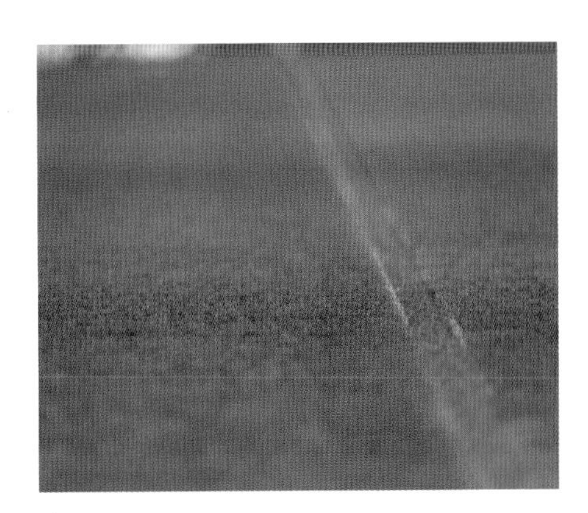

I check the forum and supporters wonder why we're watering when it's been raining. They probably think we're mad.

DAN SPARKS BRISTOL CITY HEAD GROUNDSMAN

'At Ashton Gate, there's a gentle camber to help drainage from the centre but, in all honesty, we don't need it any more. That's why, even if it rains between 9 and 12 on match day, we'd have the [integrated] sprinklers going before the match and at half-time. The soil might be damp but the leaf blade needs moistening. It just helps quicken up play. I check the forum and supporters wonder why we're watering when it's been raining. They probably think we're mad.'

Desso's projected lifespan is 10 years, but both Sparks and Robertson have known it to last up to 15 years – unlike rye grass, which is the common seed choice. (Huddersfield got 15 years out of their pioneering Desso pitch before relaying it.) The majority of top clubs strip the surface back to the Desso layer in the off-season. They then reseed, rather than using turf, because this allows for greater quality control; it takes two weeks for a good surface to appear. 'We'd do the same with our training ground,' says Robertson, 'but you have to negotiate with the coaching staff to release certain pitches. They need enough to finish the season; we need enough to begin reseeding for the off-season. Basically, you phase the new ones in so we'll still be working on some when the players return.'

This rye–Desso mix is seemingly the Holy Grail of passing football, but there are downsides. Sparks mentions that they've had problems in the past with nematodes (parasitic worms) killing the roots of the rye grass. 'It's down to a heavy sand base,' he says. 'Nematodes have free rein because nothing else really lives there.' Organic soil is now mixed with the sand to 'stimulate a positive/negative biology'. But the greatest issue is paradoxically down to that excellent drainage. 'We spend terrifying amounts on fertiliser,' says Robertson, 'feeding the pitches every 10 days. The soil is so porous, which is good for drainage but bad for holding on to nutrients.'

At West Ham's training ground, the mix of pitches includes cheaper fibre-sand construction – 'albeit still £500,000,' Robertson adds – plus two hills at 6° and 10° for fitness and rehab work. Fibre-sand is similar to Desso, but the fibres are added randomly in the sand rather than being injected deep into the ground. It stabilises the turf, but not to the same extent as Desso. 'Two-and-a-half pitches are Desso as we have one area designated for the keepers. But we're also experimenting with AirFibr. It's a French product and is a mix of cork, synthetic fibres and silica sand. Steve Braddock [head groundsman] at Arsenal introduced it to me. We've used it for our rehab area and the feedback from players has been very good. It's also holding on to nutrients much longer so we're fertilising less.'

AirFibr is just one of many innovations cranking up the quality of pitches. Some of them are simple but effective, like the Height-of-Cut Prism Gauge, which as the name suggests is a solid glass prism that enables ground staff to accurately measure blade height. In the Premier League you can't go over 30mm or you risk being fined. ('There's no minimum,' adds Robertson, 'but you wouldn't go under 20mm as it damages the grass.') Other pieces of technology are the stuff of sci-fi…

Clubs in the shade

At Premier League stadia and training grounds across the country, mobile lighting rigs are wheeled on to pitches to stimulate grass growth. The same thing happens down in the Championship at Bristol City's Ashton Gate.

'Shade is a massive problem, especially in stadia,' explains head groundsman Dan Sparks. 'For plants to photosynthesise, clearly they need light. And that's just not possible with the size of some stands.'

Or 'non possibile', as they'd say in Italy. Milan clubs AC and Inter, who share the San Siro (officially known as the Giuseppe Meazza Stadium), know more than most the repercussions of blocking out sunlight. In the build-up to the 1990 World Cup, prohibitive costs caused the two clubs and their municipal landlord to abandon the idea of building a new stadium. Instead, they adapted the San Siro, adding a

▼ The steep stands of the San Siro historically caused the Milan sides problems

▶ Cameroon beat Argentina 1–0 at the 1990 World Cup. And then the pitch problems began…

third tier to accommodate just over 87,000 supporters (since reduced to 80,018 after further renovation work completed in 2008). Eleven towers were built with reinforced concrete, four of which provide support for the huge reticular beam that supports the roofing.

Come 8 June 1990, Milan's 'temple of football' hosted the opening match of the tournament. A 67th-minute goal from François Omam-Biyik provided one of the biggest shocks in football as Cameroon beat defending champions Argentina 1–0.

All was good … but then the problems began. That third tier might have increased capacity but it also hampered photosynthesis, meaning grass simply couldn't bed in properly. By the time Barcelona had lodged an official complaint with UEFA about the state of the pitch in March 2012, after a 0–0 draw with AC Milan in their Champions League quarter-final, the San Siro pitch had been relaid 45 times since the 1990 extension.

That summer, a Desso pitch was installed, which, allied with lighting rigs, has improved the situation. Even so, UEFA's senior operations manager, Keith Dalton, commented just two weeks before the San Siro hosted the 2016 Champions League Final between Real Madrid and Atlético Madrid that the pitch 'isn't up to the level it needs to be'.

'You can't ignore light as the grass will go dormant,' says Sparks. 'Throw in the damage it gets during games and it just doesn't recover.' Bristol City aren't the highest-profile club in the country, but they have an ambitious and wealthy owner, financial services billionaire Stephen Lansdown. Lansdown has invested millions in Ashton Gate, which has seen comfort and capacity rise along with the stands.

Those rising stands are why the club have turned to Dutch company SGL (Stadium Grow Lighting), who are experts in pitch management. SGL roll out their lighting rigs all around the world, from little-known Slovak club Spartak Trnava and Argentinian outfit Estudiantes de La Plata to footballing behemoths Barcelona. The company was founded in 2001 by former rose grower Nico van Vuuren, whose horticultural background meant he understood the issues clubs were having with lack of light.

'While we have four [rigs], we'd ideally like five as they really work,' says Sparks. 'But they're not cheap, costing £100,000 each. On top of that you have electricity usage, which comes in at around £5,000 per rig per month.'

Sparks then shows me an A4 sheet featuring a diagram of the Ashton Gate pitch. Within the dimensions is a grid of four columns and three rows with an extra square on the top row to form an upside-down L-shape. The numbers '1', '2' and '3' are marked on different squares of the grid. It looks like a slightly restricted and elongated game of Sudoku, but it's actually Sparks' plan for optimising use of the rigs.

'I have a light meter that measures sunlight. Levels are relatively low at this time of year because the sun's below the southern hemisphere. This is the plan to position the lighting rigs in the big L-shaped shadow that's created as the sun rises above the South Stand and moves around the Lansdown Stand.

'The numbers correspond to different days. The squares marked "1" get the rigs on the Monday; "2" is Tuesday; "3" is Wednesday; and then it's back to "1" on Thursday… There's no number in the bottom corner, because we don't have enough rigs to provide the whole pitch with the energy grass needs to grow.'

They covered the whole pitch the previous year but the rigs were spread too thinly, resulting in minimal growth and wasted resources. 'There's a degree of trial and error, especially as the stands are higher than they used to be.'

Has he asked Lansdown to dig slightly deeper into his Guernsey-based pockets, to buy an extra rig? (Wembley has a comprehensive set-up of 12 rigs that covers the whole pitch in two shifts.) He has, Sparks replies, but that would probably only happen if they got promoted to the Premier League. 'Currently, the board would have to assess whether that extra £100,000 should go towards a rig or towards a player…'

Major Premier League clubs like Spurs and Liverpool also have lighting rigs at their training grounds. 'We don't have them here, though,' says West Ham's Dougie Robertson during my visit to the club's Romford training ground. 'Because of light pollution. We're in the middle of a residential area. Mind you, that's not the only reason we don't have them here: a lighting rig requires 64-amp, three-phase power. I'd need either a 300m cable that could trip players up or a noisy generator. Well, neither's perfect.'

West Ham's southern location and the expertise of Robertson and his team have seemingly overcome this lack of equipment as, though we are deep into autumn, the pitches are exemplary. 'We compensate by using a tent that pumps out carbon dioxide,' he says. 'It's roughly the size of the six-yard box and you roll it around the pitch. Plants love it as it aids their recovery.'

Versatility and innovation take many forms. Down at Southampton's Staplewood training centre, nestled between the groundsman's storage unit and the first team's main training pitch, is an incongruous square of turf seemingly dropped on to the car park.

'It's a big lump of grass that's grown on a pallet. Because that position's south-facing, it enjoys quite a bit of sun,' explains Mo Gimpel, the club's head of performance science. 'It also gets wind there, which the grass needs, too. Once fully grown, the grounds folk cut it out and place it in Fraser Forster's goal as that area cuts up quickly.'

Artificial airflow

As we all learned at school, plants take in carbon dioxide from the air and emit oxygen. Like us, they need fresh air; they don't want to breathe in what they've just breathed out. However, unlike us, a plant's idea of fresh air is air containing plenty of carbon dioxide. Hence, that tent over at West Ham.

Nature's most effective method of freshening up the air is wind. Most training grounds, like Manchester United's Carrington complex, are based on huge swathes of exposed land, so lack of wind is rarely an issue. Even training facilities that evolved in housing estates, like those of West Ham and Liverpool, are only sheltered by moderate-sized homes.

Bowl-shaped stadia like Wembley really suffer [from lack of airflow]. That's why they'll use industrial-sized fans to replicate real wind.

DAN SPARKS BRISTOL CITY

It's at stadia where problems occur. Take the San Siro, whose steep stands limit wind as well as light. The modern preference for wraparound designs hasn't helped. 'Bowl-shaped stadia like Wembley really suffer,' says Bristol City's Dan Sparks. 'There's hardly any airflow. That's why they'll use industrial-sized fans to replicate real wind.'

It's the same over at the London Stadium, West Ham's new home. As the club's head groundsman, Dougie Robertson, explains, 'It was always intended to be a multi-purpose venue, but its original use was for athletics [at the 2012 Olympics]. The way it was constructed was to stop any physical air movement inside, so that when the athletes were hurdling, sprinting or throwing there'd be no wind effects to interfere with times and distances.

'That's why the stadium maintenance contractors, London 185, use fans to circulate the air. Those fans also dry out spots, which you need to do because algae loves damp, dark places. And if algae grows, grass doesn't.'

Which groundsman has got the toughest gig in UK football, I ask Robertson. 'The Millennium Stadium [renamed the Principality Stadium in 2016] is one of the trickiest because the stands are so high,' he replies. 'Lee [Evans, head groundsman] has been hammered on social media this week because of how the pitch looked when Wales played Australia at rugby at the weekend. People forget that there'd been boxing on just two weeks before. Ultimately, we can only work with what's left. Yes, technology can help grass recover, but if it's dead, it's dead.' Robertson then points at four tubs of anti-algae treatment. 'They cost £800 each. Which goes to show how much money you can save if you get airflow right.'

Yes, technology can help grass recover, but if it's dead, it's dead.

DOUGIE ROBERTSON HEAD GROUNDSMAN, WEST HAM UNITED

Designing harmony

The difficulty of designing a grass-friendly stadium that can hold 80,000 supporters has been the source of lots of business for companies like SGL and lots of headaches for grounds staff. But stadium architects and their clients are beginning to wake up to the issue of pitch maintenance.

'One of the most forward-thinking is the Emirates Stadium,' says Bristol City's Dan Sparks. 'The design maximises sunshine time. Also, when Arsenal aren't playing, they can open parts of the stadium to let wind flow in. It's really compatible with good football.'

Highbury, Arsenal's home for 93 years, had one of the finest playing surfaces in the Premier League – it certainly appealed to the fabled 'invincibles' who remained unbeaten en route to the 2003–2004 league title. With Arsène Wenger – renowned for his love of crisp passing football – heavily involved in the design of the new stadium, pitch quality was always going to be one of the key drivers.

By enforcement and design, they've created a pitch that might be even better than Highbury's. Enforcement because council planners restricted the height of the new stadium to 46m; design through numerous factors aimed at supporting Mother Nature. For example, the roof slopes inwards and its inner ring is translucent, both of which increase the amount of sunlight that shines on the pitch. In addition, the undulating roof form means gaps are created in the corners to allow air circulation and further light.

But it's Tottenham Hotspur's new 62,000-seater stadium, built within the shadows of the now flattened White Hart Lane, that's currently attracting the attention of grounds teams around the world as Spurs will trump their north London rivals by becoming the first major sports club to feature a dividing retractable pitch.

From 2019, Harry Kane, Dele Alli and co. will line up on a Desso pitch that's actually split into three 'invisible and undetectable' sections that sit in three pitch-long steel trays each weighing more than 3,000 tons. In around 25 minutes, this 'real' turf can retract under the South Stand to reveal a synthetic playing surface below. This artificial layer will be used for concerts and American football matches after chairman Daniel Levy agreed a 10-year deal with the NFL to host a minimum of two matches per season.

'It'll be impressive,' says Southampton's Mo Gimpel. 'I remember visiting Toronto a few years ago and ended up watching the Blue Jays play baseball. I had some time to kill afterwards so stayed behind. And I'm glad I did, as, within around 20 minutes, the stands parted and sections of the playing surface were taken away, all to transform it into an American football stadium. It was incredible.'

The Toronto Argonauts may not play at the Rogers Centre any more, but one fact highlights the future of funding these huge stadia: the highest attendance at the Rogers Centre was nothing to do with baseball or American football but came in 2002 when Triple H and The Rock were in town for WrestleMania X8!

Maximising venue revenue

'The pressure on clubs to make money is huge,' says West Ham's Dougie Robertson. 'That's why they host concerts and other events. It's all about secondary revenue. And that makes financial sense because these stadia cost hundreds of millions of pounds. The problem is, it's all to the detriment of the pitch.'

▼ Clubs maximise revenues with concerts. Here, Leicester's King Power Stadium hosts Kasabian

You only need look at the past couple of summers for high-profile examples of this venue monetisation. In 2016, with Leicester City still in blissful shock from their Premier League triumph, their King Power home welcomed hometown rock band Kasabian, while June 2017 saw thousands of fans let Robbie Williams entertain them at Manchester City's Etihad Stadium. But feel for the London Stadium's grounds staff, as in summer 2017 not only did the venue rock to Guns N'Roses, Depeche Mode and that man Robbie Williams again, it also hosted the World Athletics Championships.

Or take Wembley – England's national stadium stages 70 events a year including 40 major football, American football and rugby league matches, 20 corporate days and 10–15 'non-pitch events' (mostly concerts).

And then there are clubs like Huddersfield, Swansea and Bristol City whose stadia host football and rugby matches all year round. 'We have at least one game a week at Ashton Gate,' says Bristol City's Dan Sparks. 'Last year we staged 68 games, 67 the year before. Not only is it City and Bristol rugby club but also other matches for "community engagement". We have to play a certain number of under-23 games here, too, to comply with FA rules.'

But the pitch that gives its groundspeople the most debilitating migraine is surely at Rodney Parade, home not only to League Two football club Newport County but also to *two* rugby union sides – Newport RFC and Newport Gwent Dragons. In the 2016–2017 season, the surface resembled a wasteland with County having to postpone matches because of waterlogging. That's why, come the 2017 close season, Newport opted for a newer version of Desso called PlayMaster. This is similar to Desso GrassMaster but has real grass fibres interwoven with the artificial blades, which is then rolled out like a carpet. It's also used by Ajax and certainly impressed the local Welsh media, who ran news of the development under the headline: 'From swamp to billiards table'.

Now that pitch technology has evolved to take some of the strain out of upkeep, certain groundsmen have shown their more playful side through unorthodox mowing patterns. In May 2016 Leicester City's head groundsman, John Ledwidge, hit the headlines after he cut the club's crest into the centre circle of the King Power pitch to celebrate the Foxes' unlikely Premier League triumph. Fun for the fans but frowned upon by the Premier League, who promptly aligned design rules with UEFA regulations that insist on 'straight lines, across the width of the pitch, perpendicular to the touchline'.

The fact that clubs have reached a stage where they can decorate their pitches with intricate designs shows how far football has come since the days of Ronnie Radford. Undersoil heating (see box), daily mowing to reduce 'blade shock' and encourage growth, vacuums to consume grass debris and prevent algae ... the list of innovations to stimulate pitch perfection is staggering. And they're needed, as the Premier League's global audience demands the most attractive football possible, which in turn demands a reliable playing surface.

Undersoil heating

Undersoil heating has been the not-so-secret weapon of groundskeepers up and down the country since Everton installed the UK's first in 1958. They dug up their Goodison pitch and circulated 30km of electric wire, which cost £16,000 (equivalent to around £400,000 today). Unfortunately, drainage issues caused by melting snow and frost resulted in the pitch being relaid just two years later.

In the 1970s clubs started to install large heated pipes under their pitches and the technology has changed very little since. There has been the occasional more affordable solution over the years. For example, in 1971 Leicester City paid £5,000 (£70,000 in today's money) for a giant tent called a polysphere to be installed over the Filbert Street pitch. It was kept afloat by huge blowers; in fact, it was so 'afloat' that the players could train under it. It didn't catch on.

Nowadays, the undersoil heating is near omnipresent in the Premier League as well as at more monied lower-league clubs. 'We have undersoil heating at Ashton Gate,' says Bristol City head groundsman Dan Sparks. 'Essentially, pipes sit on top of 10cm of gravel to drain water, though they're far from boiling hot: around 10°C.

'In all honesty, we don't use it very often. If you heat too much from underneath it dries out roots, which affects grass growth. It's only there to deal with surface frost. We have a pitch-temperature sensor; when it hits 3°C, the heating clicks on.'

Unlike in the Bundesliga, undersoil heating isn't actually mandatory in the Premier League. There is a rather vague rule stating that clubs must use an 'adequate system of pitch protection to the reasonable satisfaction of the board'. The English Football League's rules are similarly opaque.

West Ham's Dougie Robertson says that their training pitches are independently tested four times a year for factors like consistency of bounce and drainage, while there's even greater scrutiny at the stadia. 'All teams in the Premier League have their pitches marked every week,' says Dan Prest of the Institute of Groundsmanship. 'The referee and appointed match delegate mark them out of 10. Around springtime, the six clubs with the most marks will have a more detailed assessment based on the following factors: whether or not a club has had more fixtures due to playing in Europe, a comparison of available resources and a series of tests that are consistent at each venue.

'From this a top three is selected before an overall winner is chosen. It's fair to say the competition is healthy and the overall quality across the Premier League is revered throughout the football world.' For the record, the champions for 2017–2018 were Watford, who took Manchester United's title. Before that it was Aston Villa.

Of course, it's all very well rolling out the perfect foundation for peak performance, but if the players' contact point is wanting, the grounds team's efforts are potentially in vain.

Tales from the boot room

Arsenal's 1–0 home defeat to Chelsea in January 2016 was instantly forgettable, a Diego Costa goal following Per Mertesacker's dismissal offering two moments of excitement on a drab midwinter's day. But that's not how Adidas saw it, for this insipid London derby provided the backdrop for one of their sponsored stars, Mesut Özil, to become the first player in Premier League history to take to the pitch in laceless boots.

'My whole career I've tried to minimise the impact of laces on my strike and ball control,' Özil cooed at the time. 'I revealed to Adidas in 2014 that in the changing room I knot the laces over and over again and then tuck in the ends. That way they do not interfere with my touch. So when they came to me last year with the laceless boots, it was like they had created my dream product. I cannot wait to wear them on the pitch.'

Adidas claimed that the German's ACE 16+ PureControl boots, and those of Barcelona's Ivan Rakitic and Bayern Munich's James Rodriguez, locked in the foot as comfortably and securely as laced boots, thanks to three key points of stability. A thermoplastic cage provides structure at the mid-foot, with further support offered by the knitted upper and what Adidas call 'an internal knitted techfit locking system'.

Özil's vivid green boots were the latest entry into an increasingly competitive – and lucrative – world. Football is the most marketable sport on the planet with manufacturers maximising their star players' power as influencers via social media. When Nike release a new version of their Mercurials, they tap into Cristiano Ronaldo's 170 million Instagram and Twitter followers (albeit clearly there will be some double counting in that total). Adidas do the same with Lionel Messi's Nemeziz shoes and his 85 million Instagram disciples (Messi's not on Twitter). When Paris Saint-Germain's Qatari owners paid £200 million for Neymar, they knew they were buying not just a special player but the attention of his 120 million social media followers.

Between them, Nike and Adidas are worth upwards of £20 billion, of which the football-boot sector makes up a significant proportion. Like rivals such as Puma, Under Armour and Asics, they make a football boot to suit every pitch and every weather condition. And by the look of it, the majority of them are hanging up in West Ham's boot room. Each player's column of hooks holds 12 to 15 pairs, providing a kaleidoscope of colours to distract from the mildly musty aroma.

'There are loads of criteria to explain why a player might choose a certain pair over another,' says West Ham's Dougie Robertson. 'He might look for different stud set-ups depending on the firmness of the pitch or it could be to do with the position they play. [On-loan goalkeeper] Joe Hart, for example, will have slightly longer studs than Michail Antonio, who's a lively, fast-running winger. He doesn't need quite the same traction as Hart, whose slip could cost us a goal.'

It's a similar scene over at Swansea City's new £10 million Fairwood training base. 'Players here have around 10 pairs of boots each,' says the club's kit man, Michael

▲ Mesut Özil debuts his vivid-green Adidas ACE16+ PureControl boots against Chelsea in 2016

Eames, before elaborating on traction choice. 'I'd say there's a 50-50 split between studs and moulds (though some of those studs also feature blades). Some players will wear moulds no matter what the weather and some players will always choose studs. Others will switch between the two depending on the weather, pitch condition or just the way they feel that day.

'When we travel to away games, most players will take at least one pair of each. Some players travel with two pairs of boots (I always make sure a player has a spare pair just in case one splits during the match); other players will take four or five pairs. The players who take the most boots usually wear one pair for the warm-up, then another for the game.'

The personal touch

Football boots at the elite level are often single-use items, and it's this disposability that's to blame for the condensation on my glasses back at West Ham's Romford training ground. 'That's the steamer,' explains Robertson, pointing towards the silver tank that has just had its lid removed. 'The players put their boots in there before training or a match. It softens the boots for a better, comfier fit.'

Shin pads uncovered

THE MOST EASILY OVERLOOKED PIECE OF FOOTBALL EQUIPMENT HAS NOT BEEN IGNORED BY PRODUCT SCIENTISTS

German forward Thomas Müller has the word 'Poldi' printed on his to pay tribute to former international teammate Lukas Podolski; Cristiano Ronaldo's are adorned with pictures of family – as are those of former Real Madrid teammate Marcelo. Yes, we're talking shin pads. Whereas the original shin pad was no more than a cut-down cricket pad, kept in place by leather straps, today's footballers protect their livelihood with something a touch more sophisticated.

Most shin pads have a hard outer surface constructed from a thermoplastic material like polypropylene. The inner part is often made from ethylene vinyl acetate (EVA) or similar foam-type material for both protection and comfort. How they work is pretty simple. You get kicked and the pads spread the impact load over a wide surface area to reduce the effect before it penetrates through to your shin.

Over the past few years, however, carbon-fibre shin pads have begun to appear on the market. And with good reason… Research presented by Dr Yaşar Tatar and colleagues at Marmara University, Istanbul, in the *Journal of Sports Science and Medicine* in 2014 pitted three commercially available polypropylene shin pads, including the Adidas Predator and UCL editions, against two custom-made carbon-fibre versions. The testing set-up consisted of a prosthetic foot attached to a pendulum, which swung at an artificial tibia loaded with sensors, protected by a shin pad and wrapped in material to replicate a football sock. The aim was to see which type of pad dispersed force most effectively.

Each shin pad took a kicking at low force and high force with the sensors recording the amount of energy that drove through the pad and into the artificial tibia. And what did the study find? 'Carbon-fibre shin pads provided better protection at both levels of impact,' the authors concluded.

◄ Shin pads often comprise a hard outer surface constructed from polypropylene

Many professional players wear a new pair each match, so the boots never have a chance to 'bed in'. That's why steam machines like West Ham's are universal at clubs around the world. Another commonly found tool is a shoe-stretching machine that can give the player a half-size increase.

All these boots and fitting devices are designed to give the player that bit more feel. Proven, established, but hardly cutting-edge. Where are the bespoke, hand-crafted boots designed to fit the contours of the Premier League's most highly paid feet? 'Some of the players will have custom-made insoles to suit their shape of boot and foot and we'll bring in specialists to analyse the players' running movement and how they land,' says Robertson. 'But none of them have totally custom-made shoes.'

We have three high-profile European players for whom we build boots to those players' specs ... Our other players wear boots straight out of the box that you could wear at retail.

KYLE ALBRECHT

UNDER ARMOUR SPORTS MARKETING MANAGER

Trying to discover from Nike and Adidas whether their marquee players such as Ronaldo and Messi have fully customised shoes is akin to extracting a sword from a particularly stubborn stone. This is because both companies remain adamant that consumers are wearing the exact same shoes built to the exact same specifications as their heroes. However, it's hard to believe that in this day and age world-class players don't have integral fitting factors like lasts designed for their personal dimensions and boots made from materials of their choosing. It happens in athletics, which is a financial minnow compared to football's great white shark. And it happens at Under Armour.

'We have three high-profile European players for whom we build boots to those players' specs and with materials that are customised for that particular player,' says the company's sports marketing manager, Kyle Albrecht. 'We use certain elements of the boots we sell, while accommodating all the players' requests for weight, outsole design and width. Our other players wear boots straight out of the box that you could wear at retail.'

Whether boots are customised or not, technology is at the heart of the industry. Over in Germany, Adidas have invested in a cutting-edge laboratory that uses all manner of innovations in search of the perfect football boot, including slow-motion cameras to ascertain how specific parts of a shoe perform when a player kicks the ball or changes direction. Messi's Nemeziz boots, made from a wrap-like assembly of torsion tapes, were the result of this kind of analysis. '[Nemeziz] was inspired by the way ballerinas wrap their feet and boxers wrap their hands,' product manager, Philipp Hagel, told *Men's Health* magazine in 2017. 'To test durability, we put prototypes in a machine that flexed them 100 times per minute. Then we did a minimum of 80 hours' testing.'

Another striking feature of the Adidas lab is an automated flywheel with an artificial foot named 'Roboleg'. It's capable of kicking balls at up to 160km/h – around 40km/h faster than professional players – and is linked to 16 Hawk-Eye cameras that measure spin rate and speed of shots, so that researchers can test both boots and balls (more on which later).

Speed, control and power

Competition in the lucrative football boot market is fierce and the major manufacturers vie with each other to launch 'game-changing' knitted uppers, synthetic soles or lacing systems. Sexy terminology confuses choice even further. Thankfully, there are independent experts like Katrine Okholm Kryger. Kryger is just completing her PhD in football boot design at Loughborough University and, as they say, what she doesn't know about football boots isn't worth knowing.

'Broadly speaking, there are four criteria of football boots,' she explains. 'There's the "heritage" boot, which includes models like the Adidas Copa Mundial and Puma Kings. And then there are three sectors that are performance based. First you have lightweight "speed" boots [like the Adidas X 17+ PureSpeed]. Then there's "touch control" for dribbling and passing [Adidas Ace X 17+ PureControl]. Finally, you have the "power" boot for optimum kicking [Puma evoPower].'

Each manufacturer conjures up catchy monikers, but, essentially, the shoes will fall into one of these groups … and then find their way on to the feet of sponsored players, who often make their choice according to the demands of their position. For example, the evoPower is the choice of strikers Olivier Giroud and Mario Balotelli, for whom shot velocity could make the difference between hitting the net and warming the bench. Midfield linchpin Paul Pogba slips into those PureControl pumps, as they're designed for greater feel and improved touch. Then there's the speed boots. Cue reputedly the fastest player in world football – Gareth Bale. 'I need the fastest boots in the game, always,' Bale tweeted in September 2017. 'Love this new colour! @adidasfootball #X17'. In other words, the Welsh whirlwind adored his X 17+ Pure Speeds!

It's certainly a good story, but does science match the hyperbole? 'That's really what I'm trying to answer in my PhD,' says Kryger. 'I wanted to examine optimum football boot design as I discovered that there's very little good-quality research out there. Hence, people are claiming things that aren't actually measured. My main aim was to develop a test protocol for each performance sector.

'Let's start with the power boots,' she continues. 'In general, these boots feature a touch more padding in the upper. So we had subjects target the top right-hand corner and strike the ball as hard as they could. We then had them do the same with boots featuring minimal padding. In each case we measured shot velocity and accuracy. What did we discover? The shots taken with the padded boots were slightly more powerful, but slightly less accurate.'

Kryger believes that the reasoning behind the extra-thick layer of padding on the upper of power boots comes from studies that show a relationship between a player's perceived pain and how hard they think they're kicking the ball. 'So the more uncomfortable and painful the shot, the weaker the player perceives the shot to be,' she adds. 'But that doesn't necessarily match the measured reality. In fact, kicking barefoot has been shown consistently to produce the most powerful shots.'

That barefoot nirvana stretches to speed boots as their USP is lightness. Whereas a 'normal' size eight boot weighs around 250g, speed boots come in around the 160g mark. The theory is that the lighter the shoe, the faster the player can run. This ties in with the concept of distal weight and its influence on run economy.

The idea is that the same weight can have a different impact on running performance depending on how far away it is from the fulcrum of run motion at the hips. One study showed that athletes sprinting with 3.5kg of weight attached to their ankles burned 24 per cent more energy than when they ran at the same pace over the same distance with the same weight attached to their hips. One theory of why the Kalenjin people of Kenya produce so many good marathon runners is that they have skinny ankles and calves.

The problem is, the weight of the boot – and specifically that 100g saving – is such a small percentage of the player-and-kit combo that any speed or energy impact will be minimal at best. Take Andrés Iniesta, whose playing weight is around 69kg. Add in kit and boots and round things up to 70kg. Or 70,000g. If Iniesta's pair of boots saves 200g over a traditional model, that's less than 0.003 per cent of his entire mass.

Yes, the distal weight effect will have a slight impact, but even the princess-and-the-pea elite footballer will barely notice any performance uplift. That percentage of mass is even lower for a heavier footballer like Virgil van Dijk. Then again, if there's a placebo effect, arguably the research, investment and marketing have done their job. But in the case of dainty speed boots, that weight-saving could come at a professional cost: the dreaded broken metatarsal.

Metatarsal misery

Manchester City's Gabriel Jesus missed a significant chunk of the 2016–2017 season after fracturing a metatarsal bone in his right foot against Bournemouth. But it was David Beckham who thrust these long mid-foot bones into the English consciousness in 2002. A poor tackle from Deportivo de La Coruña's Aldo Duscher broke the second metatarsal in Beckham's left foot, putting him in a race to recover for that year's World Cup. Thankfully, the skills of Manchester United's medical staff and time spent sleeping in an oxygen tent saw Beckham able to play all five of England's games in Japan and South Korea.

History repeated itself four years later as the nation prayed that talisman Wayne Rooney would overcome fracturing the fourth metatarsal in his right foot in time for the Germany World Cup. He missed the first game, came off the bench in the win over Trinidad and Tobago, but ended up being sent off for stamping on Portugal's Ricardo Carvalho in the quarter-final.

Broken metatarsals are a relatively new footballing phenomenon, certainly to the wider media. Critics argue that manufacturers have sacrificed protection for style and sleekness in a bid to sell more boots. That wafer-thin knit hardly looks like much of a barrier to blades or studs, but, says Kryger, it's fit rather than impact that is the big concern.

'There are no real studies into football and metatarsal injuries so I don't have research to back this up, but footballers always wear boots very tight,' she says. 'This, it's presumed, adds feel for the ball. You then have localised pressure from studs or blades. Studs are normally located beneath the first and fifth metatarsals, which are both prone to stress fractures. With less weight and protection, you could surmise localised pressure from the studs is high in these sports. Running can cause similar issues and the shoes aren't anywhere near as tight or pressured. That's why, for me, people are focusing on the wrong thing – comfort is key.'

'We've joked that in the future if studs could be stuck to your feet, that would be the ultimate,' laughs Under Armour's senior director of global football, Antonio Zea. 'What's so interesting is that football players are taught that to enjoy good touch on the ball, they need to wear their boots super-tight, which leaves their feet horribly distorted! For me, the future must be more about the individual.'

Thanks to advances in 3D printing, this proposed move towards customisation is not as dreamily ambitious as it might seem. Which may be good news for footballers of non-Caucasian heritage. 'One of the fit problems in the football boot market is down to the last being shaped on a Caucasian male foot,' says Kryger, 'but

Anatomy of a football

THE HUMBLE PIG'S BLADDER HAS COME A LONG WAY

Nike has been the official ball supplier to the Premier League since the 2000–2001 season. Footballs are big business, so major brands like Nike and Adidas invest huge sums of money developing their products.

The innovations and technology cited above are a far cry from the early 19th century, when footballs consisted of inflated pig, ox or bullock bladders inside leather cases with the seams tied together with boot laces. Rubber bladders came in from 1862, but the basic method of construction remained the same through most of the 20th century.

Leather balls were notoriously heavy, and could double in weight in wet conditions as the leather and laces would absorb the rain; in fact, they were so heavy that players used to report

▲ Leather balls were notoriously heavy and could double in weight when it rained

constant headaches. Former England striker Jeff Astle, who played between 1959 and 1977 and was famous for his heading ability, used to say that heading a football was like heading 'a bag of bricks'. He died of a degenerative brain disease in 2002 at the age of just 59. At his inquest the coroner ruled that Astle's condition had been caused by heading heavy leather balls during his career.

Though the first synthetic balls were developed in the 1960s, the game continued to use leather versions to some degree until the 1980s. Nowadays, the humble football is packed with technology designed to provide a faster, safer, more exciting game. (And arguably to sell more footballs!)

▲ The mass-produced modern-day ball (below) has come a long way since individual hand-stitched leather numbers (above)

Secondary and tertiary colours serve to catch a player's eye and help them to identify spin, speed and trajectory as quickly as possible.

A larger striking surface has been created across the ball by reducing traditional 12-panel construction to just four. Fewer panels means the number of seams has been reduced by 40 per cent, eliminating hard spots, opening up the ball and creating a more pronounced sweetspot.

Almost imperceptible grooves and 3D ink further improves aerodynamics and the feel of the ball, helping it move through the air.

Proprietary outer material is coated in Nike's All Conditions Control (ACC) treatment. Originally used on boots, it maintains consistent friction between ball and foot in wet weather – useful when, according to Nike, the average Premier League season sees 125 days of rainfall.

Construction is a latex bladder enveloped in proprietary materials to maintain consistent air pressure and shape.

depending on where you're from, foot shapes vary a lot. For example, Africans tend to have wider feet, while Asians have slimmer feet. I've heard stories of African players wearing much bigger boots than they would normally need. They might be a size nine and wearing size 12 because they need the width.'

Under Armour's current solution to this problem is a synthetic material they call 'ClutchFit', which uses auxetic technology, meaning that it flexes when stretched. 'When a player stretches the material, instead of narrowing in the centre, it widens, which helps fit this 2D material over a 3D shape – the complex combination of curves in your foot,' Zea explains. 'When you play football, you put your boots under high stress, so you want a perfect fit; you don't want tight spots. Auxetics gives you a good fit under stress.'

Studs vs blades

Of course, it's not all about what wraps around the foot – what lies beneath is vital, too, especially when it comes to traction. In the 1954 World Cup Final in Bern, played on a quagmire of a pitch at the Wankdorf Stadium, stud selection was arguably more decisive than team selection. For the first time, West Germany played in the newly developed longer screw-in studs, while Hungary, unbeaten in four years, stuck with studs of regular length. After going behind 2–0 in just eight minutes, the Germans

◀ The outsole is designed for traction but not so much that the player can't safely twist and turn

struck three times to take an unlikely victory and one, industry experts argued, that was helped by the greater traction their longer studs gave them.

However, the humble outsole underwent its most dramatic change in the mid-1990s when Adidas released their 'traxion' technology – aka blades. These plastic tracks angled at different points aimed to improve traction on the turn, as well on drier pitches. Since then, myriad blade and stud designs have gripped professional footballers, but provoked mixed reactions.

While many pros play in blades only, some aren't so keen. In 2005, Sir Alex Ferguson banned them from Manchester United, convinced that they were responsible for Roy Keane's broken foot. 'The shot of the injury shows the shape of a bladed stud on Roy's foot,' Ferguson said at the time. 'The bladed studs are a danger.'

Wayne Rooney has twice captured headlines for extreme lacerations of the thigh and head caused by blades, while in 2002 Burnley striker Andy Payton was forced to retire after a blade impact left him needing 38 stitches in his leg. 'It was like a carving knife slitting it open' was his graphic account of the incident. Then again, studs are hardly innocuous, especially when they're made out of metal, or when cheaper plastic compounds fray around the edges.

In 2010 Professor Rami Abboud of Dundee University's Institute of Motion Analysis and Research led a study that looked into the relative merits of studs and blades. Following a series of tests involving amateur players running in straight lines and cutting in at an angle, his team found in favour of the traditional stud.

Abboud concluded that studs help to distribute the load on the footballer's feet, while the bladed design increased pressure on the metatarsals. However, the fact that only two designs were used and the test was played out on artificial turf, rather than the softer Desso mix, meant that the results were quickly dismissed by blade proponents.

'It's hard to say which is better,' says PhD boot design student Katrine Okholm Kryger. 'If your studs are too long, you won't be able to change your angles quickly and you'll get stuck in the grass. That can lead to ligament injuries around the joints. Then again, while blades are good for accelerating and decelerating, I don't feel they're great when cutting sideways; they're not optimal for multi-directional movement, which is one of their selling points.' There's also talk that the combination of blade design and the more compact grass of today's pitches has contributed to a rise in anterior cruciate ligament injuries.

'Ultimately, it needs a bespoke solution down to the individual player, surface and conditions,' Kryger concludes diplomatically.

From Craig Johnston's original 1994 Predator design for Adidas, which integrated rubber ridges theoretically to aid control, to today's Nike Hypervenom and its offset lacing designed to create a greater ball-striking area, separating what sells from what works is a tough ask. Ultimately, you won't transfer from Torquay United to Barcelona off the back of a bespoke pair of football boots. But as shown by Marouane

Fellaini's recent, ultimately unsuccessful lawsuit against New Balance, where the Belgian claimed 'defective boots' had an impact on his performance and caused a 'loss of enjoyment' and 'inconvenience', passions can run high in a player's relationship with his footwear.

It also goes to show how much football has changed. It's hard to imagine Nat Lofthouse taking the manufacturer of his boots to court. The modern-day elite footballer turns over the equivalent of a medium-sized business; in 2017 data specialists Sporting Intelligence calculated the average Premier League footballer's weekly wage to be £50,817. Player wealth is a symbol of football's growing global popularity. The game has finally gained a foothold in America, made significant moves in China and, of course, continues to grow in Europe, where the latest Premier League television deal is worth nearly £5 billion.

Football may have become more and more businesslike, but at its heart it's still a game. The growing fanbase turn on to watch Mbappé's rampages through the back line, Ronaldo's step-overs and Kane's cool, clinical finishing. Spectators demand to be entertained and, as suggested by the sackings of Sam Allardyce and David Moyes at the end of the 2017–2018 season, and the disgruntlement at Old Trafford during José Mourinho's reign, winning is no longer enough.

Gear and pitches have played a key role in football's flourishing. No matter how much talent is coursing through the limbs of Messi and Coutinho, they'd struggle to execute half their tricks if the Camp Nou's playing surface was anything like the pitches of the past. 'We had to play on some real mudheaps and ice rinks,' former England striker Jimmy Greaves wrote of his professional career, which ran from 1957 to 1971. 'There's no doubt the pitches had a massive influence on the way the game was played ... I even knew of players being picked for a particular pitch and weather rather than their ability. If we'd had a downpour a manager would sometimes turn to a player who could hoof it a long way rather than someone more likely to play pretty little 10-yard passes.' Now training and stadium pitches are pristine; boots are lightweight and feel like an extension of the leg; and footballs (see box on page 224) are designed to encourage smooth, slick passing.

The beautiful game's evolution has leaned heavily on a technological revolution that encompasses training, nutrition and equipment. Whatever your views on the influence science has had on football at the top level, it's clear that its impact is set to grow. You only have to see the effect of VAR at the Russia World Cup for evidence of that. But just how integrated will science and football become? What lies ahead for future internationals? It's something we look at in the final part of this book.

Science of the knuckleball

CRISTIANO RONALDO SHOULD THANK MODERN FOOTBALL DESIGN FOR HIS SPECTACULAR KNUCKLEBALL FREE-KICKS

There he stands, legs inching slightly further than shoulder-width apart, hem of shorts tucked in and up, chest puffed out, a deep inhalation through the nose, deep exhalation through the mouth, and whack – the ball rises, fires through the air before veering unpredictably to deceive the goalkeeper, and send ripples through the net and the crowd. There are few more wondrous sights in football than Ronaldo's knuckleball free-kick. Two things have made this possible: the construction of a modern football and the Portuguese star's relentless work ethic.

For years, World Cup balls were constructed of 32 leather panels stitched together. Synthetic materials took over, but the balls still had numerous stitched panels. All that changed in 2006 when Adidas launched the Teamgeist, made of just 14 panels attached together with glue rather than stitching. The number of panels continued to go down: the 2010 World Cup had the eight-panel Jabulani; the 2014 and 2018 World Cups showcased the six-panel Brazuca and Telstar 18 respectively.

What does all this have to do with the knuckleball? NASA's Ames Research Centre modelled the phenomenon and discovered that the Jabulani ball became unpredictable at speeds over 70km/h. The researchers concluded it was down to that panelling and construction, which made the ball smoother than its predecessors. The smoother the ball, the greater the drag at higher speeds.

'When a sphere is in a flow, there is a critical velocity at which the wake behind the sphere and the drag force acting on the ball sharply decrease,' researcher Caroline Cohen, who investigated the science behind the knuckleball, said in 2012. 'That asymmetry in the wake creates a sideways force resulting in the zigzagging motion.'

The surrounding air flow is different on opposite sides of the ball, and since the distribution of air pressure is constantly changing, the ball flutters. That 'critical velocity' (known as the drag crisis threshold) is around 70–80km/h; this is the speed Ronaldo must achieve to produce a knuckleball. It is also important that he doesn't apply any spin. And the smoother the sphere – the fewer seams in the ball – the more you see the knuckle effect.

The problem with the 2010 Jabulani ball was that it was too smooth, resulting in an extremely exaggerated movement that goalkeepers couldn't read at all. 'I noticed straight away that this Jabulani ball wasn't right,' Italy's goalkeeper Gianluigi Buffon told the BBC at the time. It wasn't just the keepers who were befuddled, so were the players taking the shots; the 2010 tournament actually produced the lowest goal average since 1990.

That's why more recent balls have featured an ever-so-slightly textured surface, which cuts down on drag but not enough to prevent the knuckleball effect – as long as the shot-taker is talented enough to make it happen.

EPILOGUE: FUTURE FOOTBALL

The beautiful game has moved on apace since Professor Tom Reilly, the man known as the pioneer of sports science in football, studied players' work rates back in 1976. Reilly observed First Division matches from a seat in the stand overlooking the halfway line, while recording observations on his Dictaphone. Distance was estimated in 1m units by using cues on the pitch and boundaries, based on which Professor Reilly would calculate sprints, distance run and jogging time.

These motion-analysis methods seem rudimentary now, but they were ahead of their time – certainly in England where coaches had, according to Reilly, no interest in scientific applications. It's why when he wrote the landmark book *What Research Tells the Coach About Soccer* in 1979, he published it in the United States.

'There was no point in putting it out in the UK,' Reilly would write in 1994. 'It would have only sold half-a-dozen copies. The Americans were much more interested in sports science in general and it did quite well over there.'

Now, as you've seen, even the British have embraced sports science and technology. But what about the future (presuming eSports don't banish real football to the sidelines!)?

Packing the players

One of the most immediate developments could be coaching staff making in-game decisions based on GPS data. Currently, club teams aren't allowed to make decisions based on 'live' data. At the 2018 World Cup, however, FIFA allowed electronic devices on the bench for the first time. Each team was permitted to use two devices: one for an analyst observing the match from the media centre, another for the coaches at pitchside.

FIFA stated the access would allow better communication with the analyst, but it's surely only a matter of time before managers at every club side in the world are checking their subjective view of a match against real-life empirical data to, for instance, replace a fatiguing player or make a tactical switch.

The likes of Catapult and StatSports (see chapter 1) will be in prime position to dominate this market in data-driven decisions, though a number of start-ups are already laying the foundations for a fruitful footballing future. Take Impect Coaching from Germany.

Impect Coaching's data platform is built on the concept of 'packing'. 'The point is that after a successful offensive action, there are fewer opponents between the ball and the opponent's goal than before the action,' Impect's founder, Lukas Keppler, told me at the inaugural Soccer Science conference held at Bristol City's Ashton Gate stadium in June 2018. 'These overplayed opponents can no longer defend their goal. They are therefore "packed" and taken out of the game.'

So if a player makes a successful pass that takes out five of the opposition, they'll earn five points. Currently, the record is Toni Kroos' 84 points in Germany's 7–1 demolition of Brazil in the 2014 World Cup semi-final. It sounds simplistic but the metrics go much deeper, taking into account other factors like where on the pitch the ball might be intercepted and how many opponents were left facing the subsequent attack.

The more data Impect collect, the more in-depth and accurate a profile of a player they can paint, resulting in a sort of technological Top Trumps. This has implications not only for in-game decisions – if a player's performance is way down on his normal stats, for example, it might be time to substitute them – but also as a scouting tool.

The more data Impect collect, the more in-depth and accurate a profile of a player they can paint, resulting in a sort of technological Top Trumps.

'It's already happening,' says Keppler. 'Let me give you an example. Before the 2017–2018 season, VfB Stuttgart sent us a profile of the kind of player they were looking for. They said we're after a midfielder who's very strong in intercepting the ball, covers a lot of ground throughout the game and helps to protect the defence.

'One of the players we came up with from our database was Santiago Ascacíbar, who played for Argentinian side Estudiantes. We showed the team our stats and a simple-to-use map highlighting his strengths, but also warned that he rarely goes past a player, so they must be wary that if you have two players like this in midfield, you could struggle to reach the last third.

'What did they do? Well, we overlaid the map of Ascacíbar's movements on to maps of their current midfielders and he complemented Christian Gentner perfectly. Gentner is the one who kickstarts attacks ... and so they made the transfer.' The pair played together 44 times in 2017–2018 and helped Stuttgart to seventh, their highest finish in the Bundesliga for six years.

This is a perfect example of the Moneyball concept in action in football. The term comes from the 2003 Michael Lewis book *Moneyball*, which explains how the Oakland A's baseball team looked beyond the 'sexy' performance statistics like batting average and home runs when recruiting players. Instead they focused on less glamorous metrics such as on-base percentage (the proportion of times a player reaches base safely by whatever method). By doing this the A's analysts were able to uncover hidden gems who fitted the team's needs perfectly and so gain an advantage over their richer rivals. As the publication date of Lewis' book shows, this approach has been going on in stats-heavy baseball for years, but until now it has been harder to apply to the more fluid, less measurable sport of football.

Social scouting

Forensic examination of potential purchases will evolve via social media, too. Premier League players have already drawn unwanted attention for historic tweets that have left them seeking forgiveness. In 2016, then Burnley striker Andre Gray, now of Watford, apologised for homophobic tweets he wrote in 2012; Chelsea winger Kenedy, currently on loan at Newcastle, was equally contrite about xenophobic and racist Instagram posts. But clubs can learn a lot more about a player by analysing their social media presence in detail.

Communications agency Cicero dissect a player's online activity to build a picture of their life to see what kind of character they are – for example, how they've reacted to fame. Some clubs even have their own in-house teams – sometimes within their scouting departments – to perform similar research. (And certain newspapers can always be relied upon to dig dirt, as happened with Gray, in order to sell more copies.)

Cicero's head of sport, Ben Wright, told the BBC in 2017 how one Premier League club ended interest in a player after the analysis uncovered a series of sexually explicit social-media posts dating back several years, as well as angry, offensive interactions with opposition fans.

Traditionally, scouts would have studied players' standing among their peers and how they might react to the media. The future will see the evidence of Twitter and Instagram become ever more important in deciding on transfers, because even the most talented player is of no use if he or she disrupts the team.

Training by genetics

Training and talent identification will become more individualised thanks to a growing understanding of genetics. Commercial outfits like DNA Fit and Muhdo are already working with a number of 'ambassadors' to promote their DNA profiling

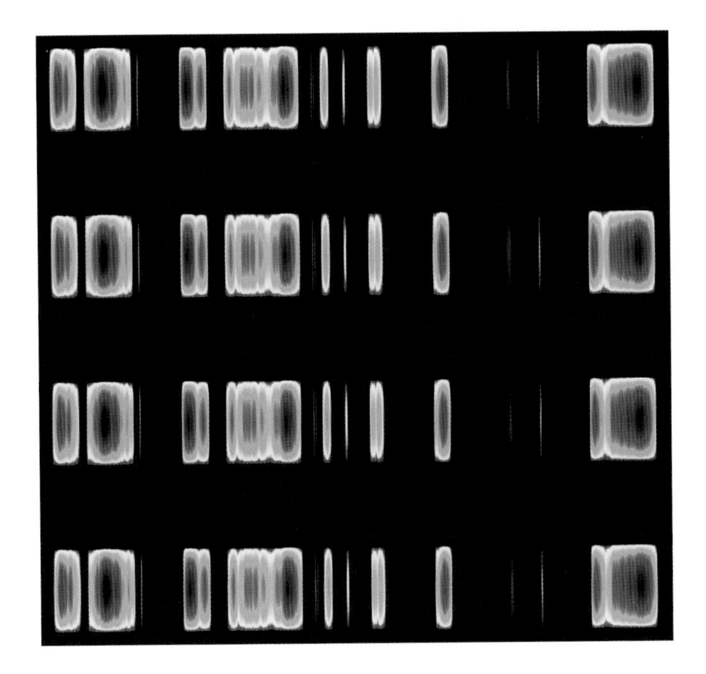

▲ DNA profiling will play an increasing role in training …

kits, with researchers believing the global market for this technology could be worth more than £8 billion by 2022.

The idea is pretty simple. The DNA testing company take a swab of the footballer's mouth, and then perform a detailed analysis of the DNA to provide the coach with 'bespoke' advice on how best to maximise that player's fitness levels.

'We test for the nucleobase pairs within a SNP (single nucleotide polymorphism) of a gene,' explains Muhdo's clinical director, Chris Collins. 'For a footballer this means that tests look at certain base characteristics that are bundled into categories such as muscle stamina, power and VO_2 max, flexibility, susceptibility to injury and so on. We also look at diet and supplement markers.' For every gene there are three possible combinations of SNP.

Based on the body of available research, testing companies like Muhdo and DNA Fit have identified certain SNP combinations within certain genes that, they say, can influence different parameters of performance. For example, within the ACE gene you can have II, DD or ID (more on which below).

Having found particular SNP combinations within your genetic make-up, the testing company can then offer customised training or nutrition advice. For example, the gene ACTN3 is associated with power; PPARA regulates fat; NRF2, respiratory capacity; and VEGF, blood vessel growth and so endurance. And then there's that ACE (angiotensin-converting enzyme) gene, which has attracted a huge amount of attention, and is involved in blood pressure control and, consequently, power and endurance.

The ACE gene first came to prominence in 1998 when Professor Hugh Montgomery and his team at University College London studied army recruits undergoing basic training. Montgomery showed that subjects with an II pairing in the ACE gene enjoyed the greatest endurance increases; those with DD the lowest. So, II was linked to endurance performance.

Simple. Not quite. Further research by noted geneticist Yannis Pitsiladis took DNA samples from 221 national-standard Kenyan athletes, 70 international-standard Kenyan athletes and 85 members of the general Kenyan population. The results showed no strong link between the II genotype of the ACE gene and endurance performance.

'The [elite Kenyan athletes'] success isn't down to favourable genetic characteristics,' argues Pitsiladis. 'Rather it appears to be the result of a favourable body type, leading to exceptional biomechanical and metabolic efficiency; chronic exposure to altitude in combination with moderate-volume, high-intensity training; and a strong psychological motivation to succeed for economic and social advancement.'

Pitsiladis feels that the field is too immature for commercialisation and that far more research is required, especially around the interaction of genes with each other. But that hasn't stopped football clubs like Barcelona pursuing a genetic advantage. In 2016, the *Daily Mail* reported that Barcelona's team doctor, Ricard Pruna, had analysed the saliva of Messi and co. to study 45 genes in an effort to work out each players' susceptibility to muscular problems. Barcelona haven't commented on how they've used Pruna's analysis, but it requires no huge leap in imagination to see a future in which major clubs make DNA testing a standard way of assessing their academy players' potential.

> The electric current makes your brain more receptive to the skill you're looking to learn.
>
> **HUGO SILVA PINTO** SPORTS PHYSICIAN

There are clear ethical arguments about this form of talent-spotting – snuffing out a 10-year-old's dreams of becoming a professional footballer because their genes aren't elite enough sounds pretty harsh, and overlooks the important influence of intangible qualities like attitude.

Stimulating skill development

That ethical tightrope also needs to be negotiated when it comes to another possible technology of the future: transcranial direct current simulation or tDCS.

A number of sporting teams, including the San Francisco Giants in baseball and skiers from the US Ski and Snowboard Association, are working with Halo Neuroscience in California to test whether stimulating the brain with electricity can improve sporting performance. Sound a touch Frankenstein?

'tDCS excites neurons in the brain, the degree of which is down to the duration of session, intensity of current and electrode placement,' explains Hugo Silva Pinto, a Portuguese physician who specialises in the procedure. 'If you apply tDCS to the motor cortex – the part of the brain involved in the planning, control and execution of voluntary movements – and then perform a task, there's evidence that you'll subsequently learn that skill faster and execute it more efficiently after.'

Essentially, the current makes your brain more receptive to the skill you're looking to learn, with Pinto citing studies that have shown a 9 per cent reduction in time taken to learn that skill. However, this greater receptiveness is temporary – like a caffeine hit, say – so the athlete would have to practise the skill very soon after a tDCS session.

Picture overleaf: Tottenham Hotspur undergo a sports-science proven warm-up, designed to improve performance and reduce the chances of injury

Halo Sport are the highest-profile company looking to turn tDCS into a commercial success. They use a device similar to a set of headphones that sends a current through to the athlete's motor cortex. Pinto, however, is sceptical of their approach. 'In my experience, the stimulation provided by Halo's tool is too low for the athletes to enjoy tDCS benefits,' he says. 'For that, you'd need to visit a clinic and use a clinically proven tDCS machine.'

That's exactly what many of Pinto's clients do, albeit primarily 'for pain management'. As for footballers using tDCS then racing outside to practise their step-overs or knuckleball free-kicks, Pinto knows of none so far. 'But as more research becomes available and players' fears are alleviated, it will only be a matter of time.'

▼ ... and DNA, more controversially, could also be used for talent identification

If all this sounds fantastical, there are plenty more ideas from a 2014 report by futurologist Ian Pearson and electronics company HTC called 'The Future of Football'. Among the technological advancements they predict are: cameras to be routinely embedded in players' kit by 2020; referees to get augmented-reality tools by 2025; and, the biggie – robot football stars to become commonplace by 2040.

Time will tell if Pearson's predictions come true, but analytics and technology certainly have the potential to transform football. Whether they do so for the better will be down to the coaches and sports scientists who are charged with implementing best practice. Mismanaged and we'll end up watching 22 automatons whose creativity has been snuffed out by numbers; managed well and tomorrow's players will be fitter and more talented and make the beautiful game more beautiful than ever.

GAZPROM
pepsi

Enjoy Responsibly
UEFA CHAMPIONS LEAGUE

INDEX

ACKNOWLEDGMENTS

Like a goal-poacher expressing their gratitude to teammates, there are countless people and clubs I'd like to thank for their time, access and insight into the world of professional football. Broadly, the sports scientists, nutritionists, exercise physiologists, coaches and support staff whose diligent work behind the scenes too often goes unnoticed.

As for the Golden Boot, that goes to the saintly staff down at Southampton, especially the diminutive Mo Gimpel. Mo's seen off numerous managers and is the sort of figure who stands tall while the walls are crumbling around him. For a man of science, he has the common touch. Thanks to Cornwall's finest footballing export, Sam Erith, who took time out from guiding the likes of Aguero, Sterling and Walker to show me arguably the most cutting-edge training facilities in the world.

Apparently, only the New York Times' Rory Smith and I have been granted interview time with Liverpool's head of nutrition, Mona Nemmer. Many thanks for the confidence you placed in me. Also, Matt Taberner at Everton for helping to kickstart the book; Tony Strudwick, football's sport-science Messiah; and Aaron Harris at Spurs, who's arguably one of Poch's secret weapons.

All the academics, universities and their subjects whose work often goes unnoticed. And to Simon Austin, editor and founder of the Training Ground Guru website. This is an excellent evidence-based newsroom that cuts through the rumour, gossip and blatant make-believe rife in professional football.

Football cap doffed to the folk at Bloomsbury Publishing. Charlotte Croft, thank you for commissioning this beast of a project. To the art and editorial team who corrected and guided where appropriate, especially Zoë Blanc who nestled in the Premier League when it came to diplomatic nudging.

Nearer to my Bristol home, fine friends Borgman and Bod (Tassell) for their professional input, plus Craig and Cuz, who like so many thousands of unsung heroes give their free time to coach young footballers, in this case my son and his AEK-Boco teammates. The peak of the footballing pyramid simply wouldn't exist without people like you. And that includes my two footballing Gods – King Eric and Captain Marvel.

Cheers for sister Lou, who always provides the calmest and clearest of team talks. My mum and dad, who nurtured my love of the beautiful game, whether decorating my bedroom with Manchester United wallpaper (standard for a Devonian!), 'coaching' me on the five-a-side pitch or heading south for a regular dose of disappointment down at Plymouth Argyle.

My beautiful daughter, Mia. I can forgive your lack of footballing know-how for your ability to percolate the finest performance-enhancing coffee. My son Harry, who can thread a pass through the eye of a needle but, more importantly, do so with simply wonderful hair. I am incredibly proud of you both. And, of course, my wife Tara. You might be a fair-weather Rovers fan but your support, encouragement and desire to be better – and to have more fun! – inspires me all-year-round. Only a few more years and you'd have managed me longer than Sir Alex's stint at Old Trafford! I love you.

PICTURE CREDITS

Pages 2–3 © GLYN KIRK/AFP/Getty Images; pages 4–5 © Matt Watson, Southampton Football Club; page 6 © Laurence Griffiths/Getty Images; page 8 © Matthias Hangst/Getty Images; page 9, 11 © Matt Watson, Southampton Football Club; page 12 © Laurence Griffiths/Getty Images; page 14 © Tony McArdle/Everton FC via Getty Images; page 15 © MICHAEL CAMPANELLA/Getty Images; page 17 © Matt Watson, Southampton Football Club; page 19 ANDY BUCHANAN/AFP/Getty Images; page 20 © Tom Szczerbowski/Getty Images; page 21 © Jean Catuffe/Getty Images; page 23 © Xavier Bonilla/NurPhoto via Getty Images; page 24 © David Ramos/Getty Images; page 26 © Chris Brunskill/Fantasista/Getty Images; page 27 © Gabriele Maltinti/Getty Images; page 29 © Photo by VI Images via Getty Images; page 30 © Warren Little/Getty Images; page 32 © Shaun Botterill/Getty Images; page 34 © Alex Grimm/Bongarts/Getty Images; page 36 © Simon Hofmann/Bongarts/Getty Images; page 40 © Darren Walsh/Chelsea FC via Getty Images; page 42 © Evren Atalay/Anadolu Agency/Getty Images; page 43 © Tottenham Hotspur FC/Tottenham Hotspur FC via Getty Images; page 45 © Adam Fradgley - AMA/WBA FC via Getty Images; page 47 © AMA/Corbis via Getty Images; page 48 © Lars Ronbog/FrontzoneSport via Getty Images; page 49 © Urbanandsport/NurPhoto via Getty Images; page 51 © DAVID HECKER/AFP/Getty Images; page 50 © Matt Watson, Southampton Football Club page 52 © John Peters/Manchester United via Getty Images; page 55 © John Peters/Manchester United via Getty Images; page 56 © Laurence Griffiths/Getty Images; page 60 © Stuart MacFarlane/Arsenal FC via Getty Images; page 64 © Dan Mullan/Getty Images; page 66 © FILIPPO MONTEFORTE/AFP/Getty Images; page 68 © Valery Sharifulin\TASS via Getty Images; page 73 © CHRISTOF STACHE/AFP/Getty Images; page 74 © Matt McNulty – Manchester City/Man City via Getty Images; page 76 © Nigel Roddis/Getty Images; page 78 © Robbie Jay Barratt - AMA/Getty Images; page 81 © Tony McArdle/Everton FC via Getty Images; page 82 © Richard Heathcote/Getty Images; page 84 © Julian Finney/Getty Images; page 85 © PAUL ELLIS/AFP/Getty Images; page 87 © Catherine Ivill - AMA/Getty Images; page 90 © Mike Hewitt/Getty Images; page 93 © Neville Williams/Aston Villa/Aston Villa FC; page 94 © Gary M Prior/Allsport via Getty Images; page 96 © Warren Little/Getty Images; page 99 © Matt Watson, Southampton Football Club page 100 © Andrew Powell/Liverpool FC via Getty Images; page 103 © Erwin Spek/Soccrates/Getty Images; page 104 © Erwin Spek/Soccrates/Getty Images; page 106 © Andrew Powell/Liverpool FC via Getty Images; page 109 © Alexander Hassenstein/Bongarts/Getty Images; page 110 © FILIPPO MONTEFORTE/AFP/Getty Images; page 113 © Matthew Ashton – AMA/Getty Images; page 114 © Chris Brunskill Ltd/Getty Images; page 116 © Dave Howarth - CameraSport/CameraSport via Getty Images; page 118 © Richard Heathcote/Everton FC via Getty Images; page 119 © David Price/Arsenal FC via Getty Images; page 121 © Matthias Kern/Bongarts/Getty Images; page 125 © LEON NEAL/AFP/Getty Images; page 126 © Matt Watson, Southampton Football Club page 129 © Media for Medical/UIG via Getty Images; page 131 © Paul CHARBIT/Gamma-Rapho via Getty Images; page 136 © Marco Canoniero/LightRocket via Getty Images; page 139 © Juan Manuel Serrano Arce/Getty Images; page 140 © Mikael Sjoberg/Bloomberg via Getty Images; page 141 © Getty Images; page 142 © Michael Regan/Getty Images; page 143 © JOSEP LAGO/AFP/Getty Images; page 145 © Chris Brunskill Ltd/Getty Images; page 151 © Ross Kinnaird/Allsport via Getty Images; page 152 © Vid Ponikvar/MB Media/Getty Images; page 155 © Matthew Lewis/Getty Images; page 158 © Matthias Hangst/Getty Images; page 160 © PA Images; page 161 © Leslie Plaza Johnson/Icon Sportswire via Getty Images; page 163 © Simon Stacpoole/Offside/Getty Images; page 166 © Ben Radford /Allsport via Getty Images; page 167 © James Baylis – AMA/Getty Images; page 168 © Richard Heathcote/Getty Images; page 169 © Richard Heathcote/Getty Images; page 170 © Cooper Neill/Getty Images; page 172 © CLAUDIO REYES/AFP/Getty Images; page 173 © Shaun Botterill/Getty Images; page 176 © David Goddard/Getty Images; page 180 © IAN KINGTON/AFP/Getty Images; page 182 © Tottenham Hotspur FC/Tottenham Hotspur FC via Getty Images; page 184 © John Powell/Liverpool FC via Getty Images; page 190 © James Baylis – AMA/Getty Images; page 193 © ADRIAN DENNIS/AFP/Getty Images; page 194 © Angel Martinez/Real Madrid via Getty Images; page 196 © Urbanandsport/NurPhoto via Getty Images; page 197 © Jasper Juinen/Getty Images; page 201 © Clive Howes/Chelsea FC via Getty Images; page 202 © Catherine Ivill/Getty Images; page 204 © Jackson/Express/Getty Images; page 205 © Matthew Peters/Manchester United via Getty Images; page 206 © Arfa Griffiths/West Ham United via Getty Images; page 207 © Matthew Ashton – AMA/Getty Images; page 208 © Marco Luzzani – Inter/Inter via Getty Images; page 209 © Mark Leech/Getty Images; page 213 © Plumb Images/Leicester City FC via Getty Images; page 217 © Tom Jenkins/Getty Images; page 218 © Jan Kruger – UEFA/UEFA via Getty Images; page 222 © ullstein bild/ullstein bild via Getty Images; page 223 (upper) © George Rinhart/Corbis via Getty Images; page 223 (lower) © Catherine Ivill/Getty Images; page 224 © AMA/Corbis via Getty Images; page 231 © Mehau Kulyk via Getty Images; page 233 © Jasper Juinen/Getty Images; pages 234–235 © VINCENT JANNINK/AFP/Getty Images. Line drawings © Shutterstock.